Powerplant

JAA ATPL Training

This book has been written and published to assist students enrolled in an approved JAA Air Transport Pilot Licence (ATPL) course in preparation for the JAA ATPL theoretical knowledge examinations. Nothing in the content of this book is to be interpreted as constituting instruction or advice relating to practical flying.

Whilst every effort has been made to ensure the accuracy of the information contained within this book, neither Jeppesen nor Atlantic Flight Training gives any warranty as to its accuracy or otherwise. Students preparing for the JAA ATPL theoretical knowledge examinations should not regard this book as a substitute for the JAA ATPL theoretical knowledge training syllabus published in the current edition of "JAR-FCL 1 Flight Crew Licensing (Aeroplanes)" (the Syllabus). The Syllabus constitutes the sole authoritative definition of the subject matter to be studied in a JAA ATPL theoretical knowledge training programme. No student should prepare for, or is entitled to enter himself/herself for, the JAA ATPL theoretical knowledge examinations without first being enrolled in a training school which has been granted approval by a JAA-authorised national aviation authority to deliver JAA ATPL training.

Contact Details:

Sales and Service Department
Jeppesen GmbH
Frankfurter Strasse 233
63263 Neu-Isenburg
Germany

Tel: ++49 (0)6102 5070
E-mail: fra-services@jeppesen.com

For further information on products and services from Jeppesen, visit our web site at:
www.jeppesen.com

© Jeppesen Sanderson Inc., 2004
All Rights Reserved
ISBN 0-88487-355-2

JA310105-000

Printed in Germany

PREFACE_____

As the world moves toward a single standard for international pilot licensing, many nations have adopted the syllabi and regulations of the "Joint Aviation Requirements-Flight Crew Licensing" (JAR-FCL), the licensing agency of the Joint Aviation Authorities (JAA).

Though training and licensing requirements of individual national aviation authorities are similar in content and scope to the JAA curriculum, individuals who wish to train for JAA licences need access to study materials which have been specifically designed to meet the requirements of the JAA licensing system. The volumes in this series aim to cover the subject matter tested in the JAA ATPL ground examinations as set forth in the ATPL training syllabus, contained in the JAA publication, "JAR-FCL 1 (Aeroplanes)".

The JAA regulations specify that all those who wish to obtain a JAA ATPL must study with a flying training organisation (FTO) which has been granted approval by a JAA-authorised national aviation authority to deliver JAA ATPL training. While the formal responsibility to prepare you for both the skill tests and the ground examinations lies with the FTO, these Jeppesen manuals will provide a comprehensive and necessary background for your formal training.

Jeppesen is acknowledged as the world's leading supplier of flight information services, and provides a full range of print and electronic flight information services, including navigation data, computerised flight planning, aviation software products, aviation weather services, maintenance information, and pilot training systems and supplies. Jeppesen counts among its customer base all US airlines and the majority of international airlines worldwide. It also serves the large general and business aviation markets. These manuals enable you to draw on Jeppesen's vast experience as an acknowledged expert in the development and publication of pilot training materials.

We at Jeppesen wish you success in your flying and training, and we are confident that your study of these manuals will be of great value in preparing for the JAA ATPL ground examinations.

The next three pages contain a list and content description of all the volumes in the ATPL series.

ATPL Series

Meteorology (JAR Ref 050)

- The Atmosphere
- Wind
- Thermodynamics
- Clouds and Fog
- Precipitation
- Air Masses and Fronts
- Pressure System
- Climatology
- Flight Hazards
- Meteorological Information

General Navigation (JAR Ref 061)

- Basics of Navigation
- Magnetism
- Compasses
- Charts
- Dead Reckoning Navigation
- In-Flight Navigation
- Inertial Navigation Systems

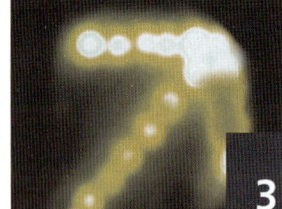

Radio Navigation (JAR Ref 062)

- Radio Aids
- Self-contained and External-Referenced Navigation Systems
- Basic Radar Principles
- Area Navigation Systems
- Basic Radio Propagation Theory

Airframes and Systems (JAR Ref 021 01)

- Fuselage
- Windows
- Wings
- Stabilising Surfaces
- Landing Gear
- Flight Controls
- Hydraulics
- Pneumatic Systems
- Air Conditioning System
- Pressurisation
- De-Ice / Anti-Ice Systems
- Fuel Systems

Powerplant (JAR Ref 021 03)

- Piston Engine
- Turbine Engine
- Engine Construction
- Engine Systems
- Auxiliary Power Unit (APU)

Electrics (JAR Ref 021 02)

- Direct Current
- Alternating Current
- Batteries
- Magnetism
- Generator / Alternator
- Semiconductors
- Circuits

Instrumentation (JAR Ref 022)

- Flight Instruments
- Automatic Flight Control Systems
- Warning and Recording Equipment
- Powerplant and System Monitoring Instruments

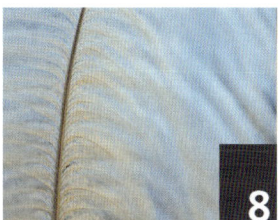

Principles of Flight (JAR Ref 080)

- Laws and Definitions
- Aerofoil Airflow
- Aeroplane Airflow
- Lift Coefficient
- Total Drag
- Ground Effect
- Stall
- C_{LMAX} Augmentation
- Lift Coefficient and Speed
- Boundary Layer
- High Speed Flight
- Stability
- Flying Controls
- Adverse Weather Conditions
- Propellers
- Operating Limitations
- Flight Mechanics

Performance (JAR Ref 032)

- Single-Engine Aeroplanes – Not certified under JAR/FAR 25 (Performance Class B)
- Multi-Engine Aeroplanes – Not certified under JAR/FAR 25 (Performance Class B)
- Aeroplanes certified under JAR/FAR 25 (Performance Class A)

Mass and Balance (JAR Ref 031)

- Definition and Terminology
- Limits
- Loading
- Centre of Gravity

Flight Planning (JAR Ref 033)

- Flight Plan for Cross-Country Flights
- ICAO ATC Flight Planning
- IFR (Airways) Flight Planning
- Jeppesen Airway Manual
- Meteorological Messages
- Point of Equal Time
- Point of Safe Return
- Medium Range Jet Transport Planning

Air Law (JAR Ref 010)

- International Agreements and Organisations
- Annex 8 – Airworthiness of Aircraft
- Annex 7 – Aircraft Nationality and Registration Marks
- Annex 1 – Licensing
- Rules of the Air
- Procedures for Air Navigation
- Air Traffic Services
- Aerodromes
- Facilitation
- Search and Rescue
- Security
- Aircraft Accident Investigation
- JAR-FCL
- National Law

Human Performance and Limitations (JAR Ref 040)

- Human Factors
- Aviation Physiology and Health Maintenance
- Aviation Psychology

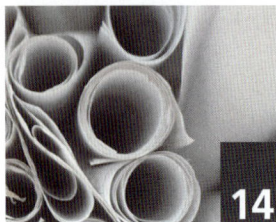

Operational Procedures (JAR Ref 070)

- Operator
- Air Operations Certificate
- Flight Operations
- Aerodrome Operating Minima
- Low Visibility Operations
- Special Operational Procedures and Hazards
- Transoceanic and Polar Flight

Communications (JAR Ref 090)

- Definitions
- General Operation Procedures
- Relevant Weather Information
- Communication Failure
- VHF Propagation
- Allocation of Frequencies
- Distress and Urgency Procedures
- Aerodrome Control
- Approach Control
- Area Control

CHAPTER 1

Piston Engine Operation and Construction

CHAPTER 2

Piston Engine Carburation

CHAPTER 3

Piston Engine Carburettors

CHAPTER 4

Piston Engine Lubrication and Cooling

CHAPTER 5

Ignition System

CHAPTER 6

Engine Operation

CHAPTER 7

Piston Engine Performance

CHAPTER 8

Piston Engine Fuel Injection

CHAPTER 9

Piston Engine Power Augmentation Systems

CHAPTER 10

Propellers

CHAPTER 11

Gas Turbine
Principles of Operation

CHAPTER 12

Gas Turbine Engines
Types of Construction

CHAPTER 13

Gas Turbine Engines
Air Inlet

CHAPTER 14

Gas Turbine Engine Compressor

CHAPTER 15

Gas Turbine Engine
Combustion Systems

CHAPTER 16

Gas Turbine Engine
Turbine

CHAPTER 17

Gas Turbine Engine
Jet Pipe

CHAPTER 18

Gas Turbine Engine
Reverse Thrust

CHAPTER 19

**Gas Turbine Engine
Internal Air System**

CHAPTER 20

**Gas Turbine Engine
Gearboxes and Lubrication Systems**

CHAPTER 21

**Gas Turbine Engine
Fuel Systems**

CHAPTER 22

Gas Turbine Engine
Starting and Ignition Systems

CHAPTER 23

Gas Turbine Engine
Electronic Engine Control

CHAPTER 24

Gas Turbine Engine
Performance

CHAPTER 25

Powerplant Operation and Monitoring

CHAPTER 26

Auxiliary Power Unit (APU)
and Ram Air Turbine (RAT)

Chapter 1
Piston Engines —
Operation and Construction

INTRODUCTION

The piston engine is an internal combustion engine working on the principle devised by Dr. Otto in 1876. The piston engine converts chemical energy in the form of petroleum fuel into mechanical energy via heat and can be termed a heat engine. The working medium is air, which is capable of changes in volume and pressure when subjected to an increase in temperature caused by the burning fuel.

The working cycle consists of four strokes of the piston: Induction, Compression, Power, and Exhaust. This is known as the four-stroke or Otto cycle. The cycle is of an intermittent nature; each stroke is distinct and separate from the others. During each cycle, the piston moves in a reciprocating motion within a tube termed a cylinder barrel. The crankshaft converts this linear motion into a rotary motion. In one four-stroke cycle, the crankshaft makes two complete revolutions — 720°.

Listed below are some of the basic terminologies required in order to understand engine operation.

➢	**Top Dead Centre (TDC)**	The position of the piston at the highest point in the cylinder.
➢	**Bottom Dead Centre (BDC)**	The position of the piston at the lowest point in the cylinder.
➢	**Stroke**	The distance between TDC and BDC.
➢	**Swept Volume**	The cylinder volume contained between TDC and BDC.
➢	**Clearance Volume**	The cylinder volume contained between the top of the cylinder and piston crown at TDC.

THE OTTO CYCLE
INDUCTION

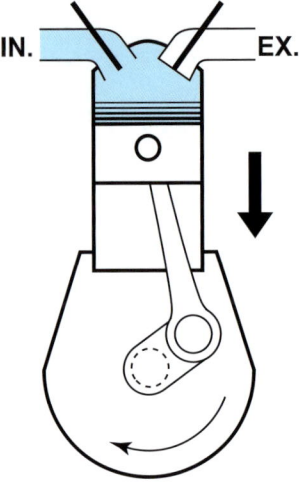

Fig. 1.1

The cycle commences with the piston at top dead centre with the opening of the inlet valve. As the piston descends, the volume of the cylinder above the piston increases, lowering the air pressure (creating suction), which is below ambient pressure. Atmospheric pressure acting on the air intake forces air through the inlet manifold, and fuel is added in the correct proportions at the carburettor. The mixture enters the cylinder through the open inlet valve.

COMPRESSION STROKE

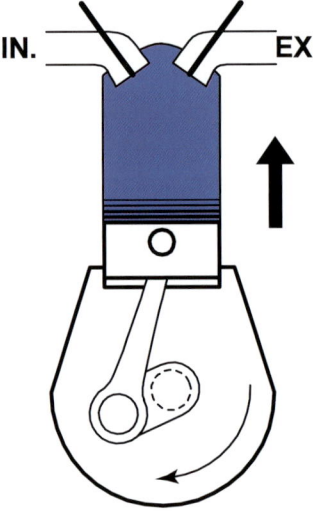

Fig. 1.2

At bottom dead centre, the inlet valve closes and the piston rises toward top dead centre with both valves closed, decreasing the cylinder volume and increasing both the pressure and temperature of the mixture. Toward the end of the compression stroke just before top dead centre, two spark plugs ignite the mixture.

POWER STROKE

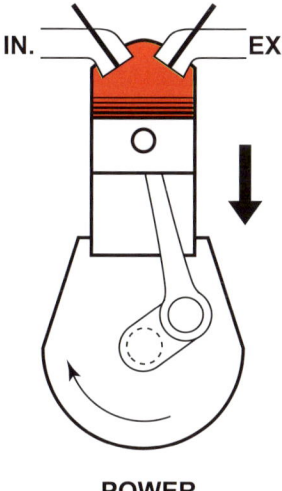

POWER

Fig. 1.3

The burning mixture expands, causing a rapid rise in pressure, which acts on the piston, forcing it downward toward bottom dead centre. The cylinder volume increases and gas pressure and temperature decrease.

EXHAUST STROKE

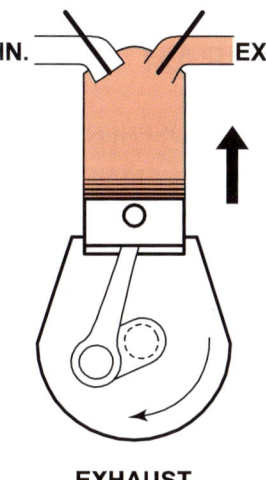

EXHAUST

Fig. 1.4

Finally, the piston rises from bottom dead centre to top dead centre with the exhaust valve open, decreasing cylinder volume and displacing the burnt gases to the atmosphere through the open exhaust valve. The process of displacing the exhaust gases is referred to as **scavenging**.

The cycle is now repeated.

INEFFECTIVE CRANK ANGLE

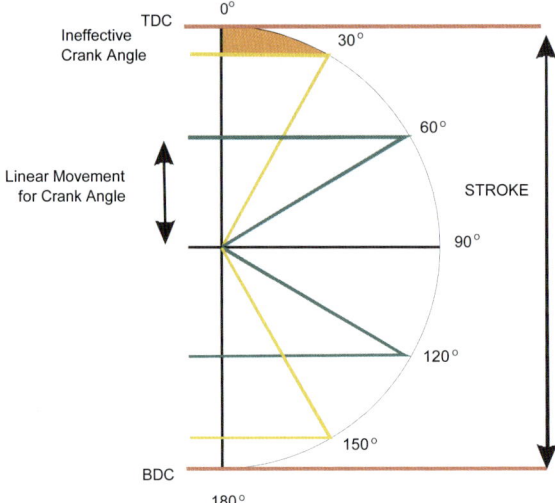

Fig. 1.5

With the valves opening and closing at dead centre positions, the engine is not efficient. To improve engine efficiency, Dr. Otto altered the valve timings to account for the time that it takes for the fuel to burn and achieve maximum pressure and for the (jerky) movement of the piston due to ineffective crank angle created by the change of rotary motion into linear motion. This is known as the **Improved Otto Cycle**.

PRESSURE VOLUME DIAGRAM

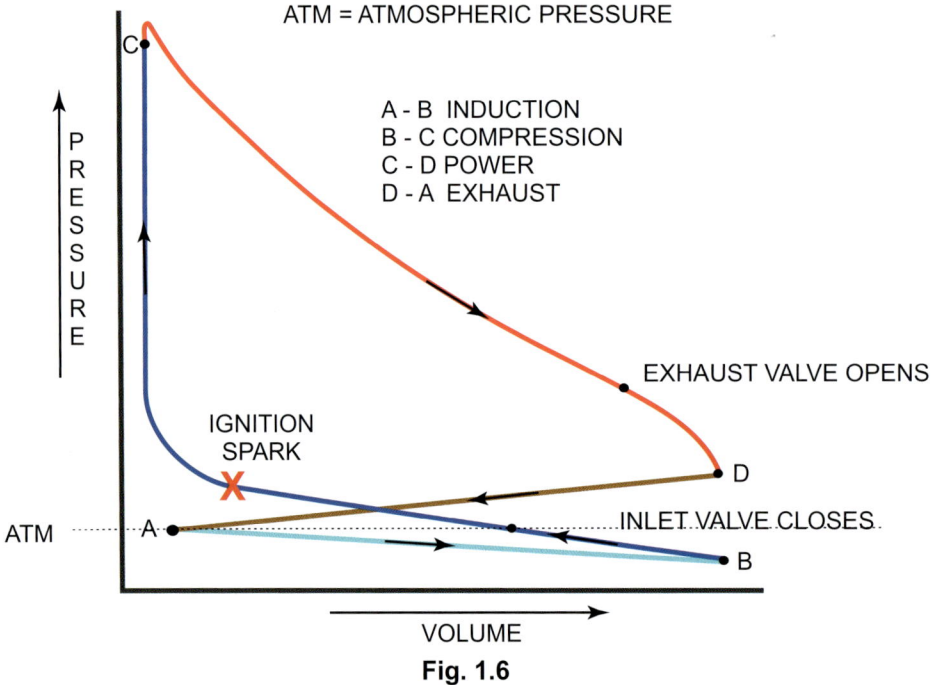

Fig. 1.6

The Ideal Pressure Volume Indicator illustrates the four-stroke cycle. Figure 1.6 shows the relationship between the pressure in the cylinder and the cylinder volume during the cycle.

VALVE TIMING

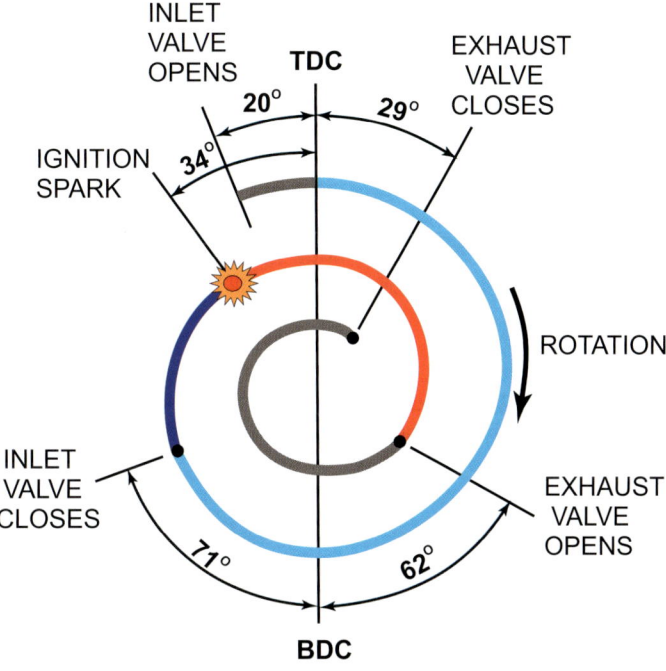

Fig. 1.7

With the valves opening and closing at dead centre positions, the engine is not efficient. Therefore, to increase engine efficiency, valve timing must be altered. Figure 1.7 illustrates the valve timing position and shows that the valves open and close either before or after the centre positions. **Lead** refers to valve operation before top dead centre and bottom dead centre positions, whilst **lag** refers to valve operation after top dead centre and bottom dead centre. The inlet valve opens before top dead centre on the exhaust stroke, whilst the exhaust valve closes after top dead centre during the induction stroke. This means that the valves are open at the same time around top dead centre. This is called **valve overlap**. Using top dead centre and bottom dead centre as a reference, the angular positions are related to crankshaft movement in degrees.

INLET VALVE (LEAD/LAG)

Inlet valve lead is the early opening of the valve during the exhaust stroke and ensures the valve is fully open at TDC. Inlet valve lag is the late closing of the valve during the compression stroke after BDC. This arrangement ensures that the valve is open for the maximum period of time and allows the maximum weight of charge to enter the cylinder.

The mixture momentum increases as the piston approaches the bottom of its stroke. It still has the energy to continue to flow into the cylinder, even after the piston has passed bottom dead centre, and the piston has travelled a small distance up the cylinder. Inlet valve closing is delayed until after bottom dead centre, when cylinder mixture pressure is nearly equal to the inlet manifold mixture pressure.

EXHAUST VALVE (LEAD/LAG)

Near the end of the power stroke, very little useful work is achieved. Opening the exhaust valve before BDC relieves the bearing load, and the residual gas pressure starts exhaust gas scavenging rapidly before the piston begins to ascend. The valve is closed late after TDC during the induction stroke and provides maximum time for scavenging. It is essential that efficient scavenging of the cylinders takes place in order that a full charge of mixture is taken in.

VALVE OVERLAP

During valve overlap, the reduced pressure in the cylinder left by the discharging exhaust gases is used to overcome the inertia of the fresh mixture in the induction system. The momentum of the outgoing exhaust gas begins pulling the fresh mixture into the cylinder before downward movement of the piston. This allows the mixture to enter the cylinder as early as possible.

The exhaust valve opens before bottom dead centre (lead). This enables the exhaust gases to scavenge from the cylinder more readily, since the gas pressure is higher than ambient. This would seem to cause a loss of pressure energy. However, vertical piston travel over 30° around top dead centre and bottom dead centre is very small and is called **ineffective crank angle**. Inlet valve lag allows time for the mixture pressure to approach the ideal, which is ambient.

IGNITION TIMING

Figure 1.7 shows that the spark igniting the mixture occurs before top dead centre to ensure that maximum pressure occurs approximately 6° to 12° after top dead centre. This ensures maximum conversion of pressure energy into mechanical energy by occurring when the piston is near the beginning of the power stroke.

To ensure that maximum pressure occurs after TDC, ignition timing ideally varies with engine speed. However, since aircraft engines operate over small rpm range, variable ignition is not necessary; therefore, light aircraft have a fixed ignition. When ignition takes place before TDC, it is **advanced**. When it takes place after TDC, it is **retarded.** It is only retarded during engine start.

After ignition, the mixture burns in a controlled fashion and the flame rate, depending on the mixture ratio, is approximately 60 - 100 ft/sec. Since maximum pressure cannot be reached until the fuel has been completely burned, ignition is required to take place well before the maximum pressure occurs. Therefore, ignition usually takes place approximately 20° - 30° before TDC.

POWER

Where the engine is in good mechanical order, the power output of a single cylinder engine depends on three factors:

> ➤ Weight of fuel/air mixture taken in
> ➤ Amount of compression of the mixture
> ➤ Number of working/power strokes per minute

The weight of mixture taken in depends on the size of the cylinder. Detonation limits the amount of compression. The strength of the materials used in engine construction limits the crankshaft speed. Since the weight of the moving parts increases out of proportion to an increase in engine size, the larger the cylinder employed, the lower the maximum safe engine speed. **Horsepower** is the measurement for power and is described below.

INDICATED HORSEPOWER (IHP)

This is the theoretical horsepower developed in the combustion chamber without reference to friction losses within the engine. It is a calculation using the formula:

$$IHP = \frac{PLANK}{33\ 000\ ft\ lb/min}$$

Where   P = Indicated mean effective pressure in psi.

L = Length of stroke in feet

A = Area of piston lead or cross-sectional area of cylinder in square inches

N = Number of power strokes per minute $\frac{rpm}{2}$

K = Number of cylinders

FRICTION HORSEPOWER (FHP)

This is the power loss due to friction and the power absorbed by the engine-driven accessories (i.e. magnetos, generators, oil pumps, etc.).

BRAKE HORSEPOWER (BHP)

This is the horsepower actually available at the propeller shaft and is always less than IHP due to FHP. BHP is normally found by practical measurement using a Prony Brake or dynamometer.

Where: **BHP = IHP – FHP**

It is impracticable to obtain much more than approximately 100 BHP per cylinder. Therefore, aircraft engines have a number of cylinders. These engines are called **multi-cylinder engines**.

CYLINDER ARRANGEMENTS

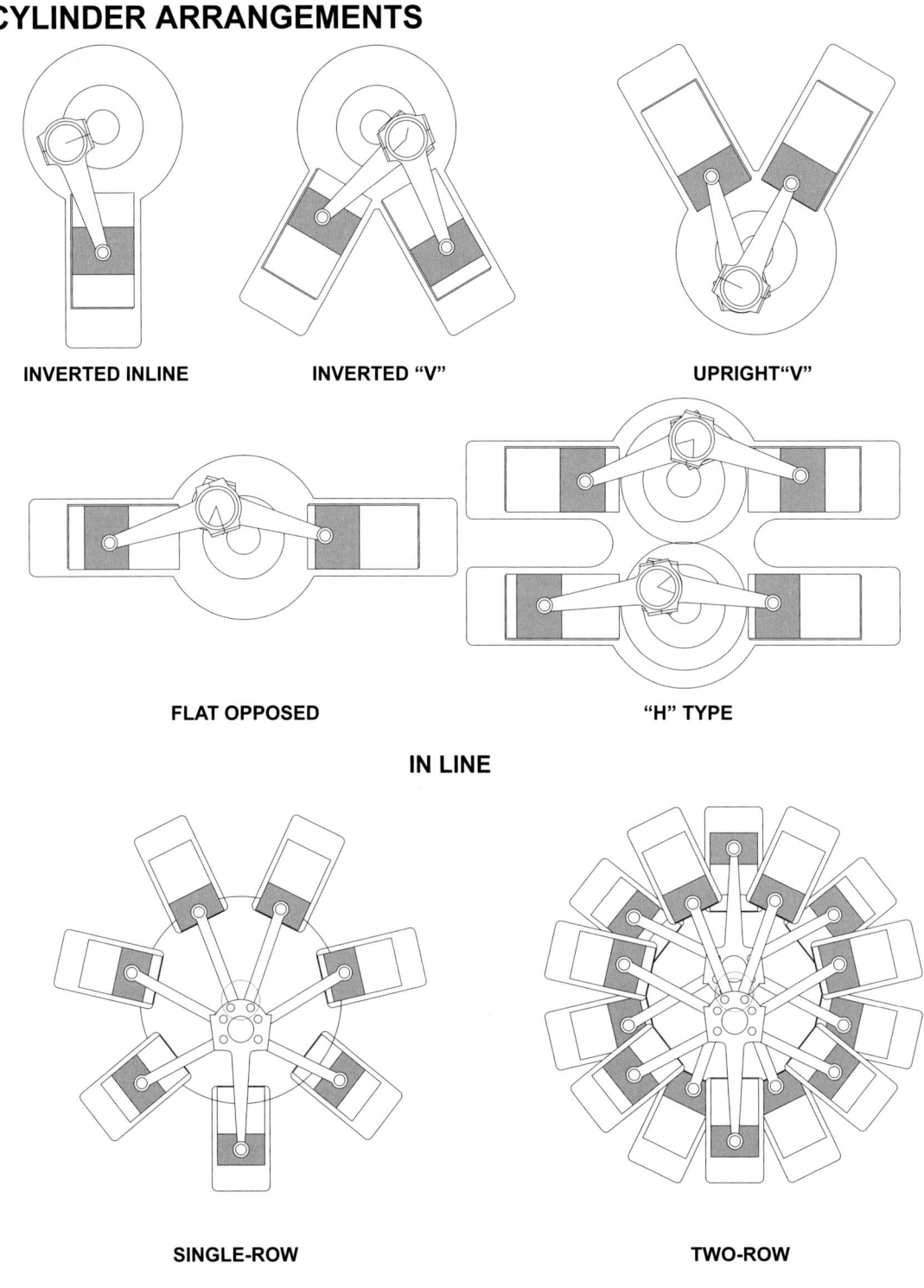

INVERTED INLINE   **INVERTED "V"**   **UPRIGHT"V"**

FLAT OPPOSED   **"H" TYPE**

IN LINE

SINGLE-ROW   **TWO-ROW**

RADIAL

Fig. 1.8

There are various cylinder arrangements that can be employed on piston engines (i.e. V, H, radial, and horizontally opposed). A brief description of the radial and horizontally opposed engines follows.

Light aircraft engines have a minimum of four cylinders, not only for more power but also to obtain smoother power. They also present a smaller frontal area, therefore reducing drag. An engine can also be classified as:

Long Stroke where the stroke is greater than the piston bore (diameter).
Oversquare or Short Engine where the stroke is less than the bore.
Square where the stroke is equal to the bore.

RADIAL

Due to the air-cooling difficulties associated with in-line engines in the early days of aviation, the radial engine was developed. In its simplest form, this arrangement has all the cylinders mounted radially in a single bank about the crankcase. This ensures that each cylinder obtains maximum cooling benefit from the aircraft forward motion and the propeller slipstream. Increased power demands resulted in an increase in the number of banks, with a maximum of four.

HORIZONTALLY OPPOSED

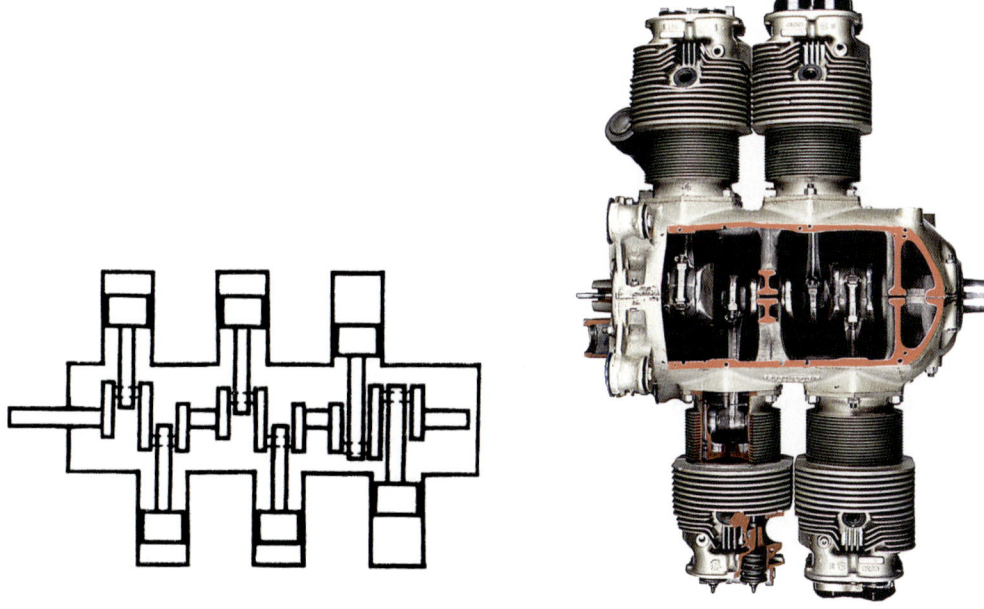

Fig. 1.9

Horizontally opposed engines have cylinders mounted on opposite sides of the crankcase. This allows the same number of cylinders to be spaced along a shorter crankshaft than in an inline engine. A modern mass-produced aircraft engine of this design is shown in figure 1.9.

ENGINE EFFICIENCIES

The efficiencies affecting engine operation are identified as follows:

THERMAL EFFICIENCY

This is the percentage of total heat generated that is converted into useful power. Should two engines produce the same horsepower but one burn less fuel than the other, the engine using less fuel converts a greater portion of the available energy into useful work. Therefore, it has a higher thermal efficiency. Thermal efficiency of piston engines is approximately 30% and can be increased by raising the compression ratio.

MECHANICAL EFFICIENCY

This is the ratio of the brake horsepower to indicated horsepower and gives the percentage of power developed that is actually delivered to the propeller.

$$\frac{BHP}{IHP} \; x \; \frac{100}{1} \%$$

VOLUMETRIC EFFICIENCY

This is the ability of an engine to fill its cylinders with air compared with their capacity for air under static conditions. A normally aspirated engine always has a volumetric efficiency of less than 100%, whereas superchargers and turbochargers permit volumetric efficiencies in excess of 100%. Various factors have a detrimental effect on volumetric efficiency:

➢ High rpm — Owing to frictional losses in the induction system and less time to feed the cylinder as rpm increases, volumetric efficiency decreases.
➢ Induction system bends, obstructions, and internal surface roughness.
➢ Throttle and venturi restrictions.

Increasing altitude reduces exhaust back pressure, resulting in better scavenging of the exhaust gas. This increases volumetric efficiency. For normally aspirated engines, maximum volumetric efficiency is achieved with the throttle fully open and the rpm as low as possible.

SPECIFIC FUEL CONSUMPTION (SFC)

This is directly related to overall engine efficiency in terms of thermal and propulsive efficiency, and is the amount of fuel burnt per hour per unit of power.

COMPRESSION RATIO

This is the ratio of the volume of an engine cylinder with the piston at BDC to the volume with the piston at TDC, and is directly related to internal cylinder pressures. The more the fuel/air mixture is compressed before ignition, the higher the pressure and temperature are after ignition. The compression ratio for piston engines is normally between 8 to 1 and 10 to 1.

The ratio expression is as follows:

$$\frac{\text{Swept Volume} + \text{Clearance Volume}}{\text{Clearance Volume}} \quad \text{or} \quad \frac{\text{Total Volume}}{\text{Clearance Volume}}$$

Also, the higher the temperature is for a given amount of fuel and air, the lower the specific fuel consumption (SFC). There is an upper limit to which the pressure and temperature in a cylinder can be raised. Exceeding this limit results in detonation of the mixture. Detonation is described in the Carburation section.

ENGINE MAJOR COMPONENT PARTS

A typical list of light aircraft engine major mechanical components is:

> Crankcase
> Crankshaft
> Connecting Rod
> Piston
> Cylinder Barrel and Head
> Valve Mechanism

CRANKCASE

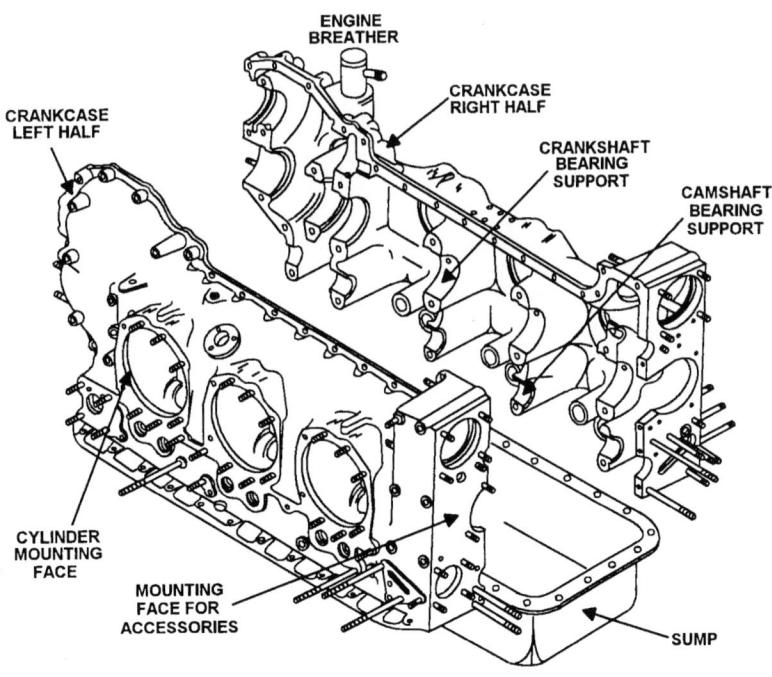

Fig. 1.10 A

This is the main engine casing and is usually made of an aluminium alloy. The housing encloses the various mechanical parts surrounding the crankshaft and contains the bearings in which the crankshaft revolves. Oil passages and galleries are drilled in certain areas to supply lubrication to the bearings and moving parts. It also contains the oil sump and forms an oil tight chamber. The crankcase provides a mounting face for the numerous accessories, such as generators and pumps, and supports the engine in the airframe. To ensure that internal pressures are approximately equal to the surrounding atmospheric pressure, a crankcase breather is fitted. Refer to figure 1.10 A. Figure 1.10 B shows the cam shaft positioned on one half of the crankcase.

Fig. 1.10 B

CRANKSHAFT

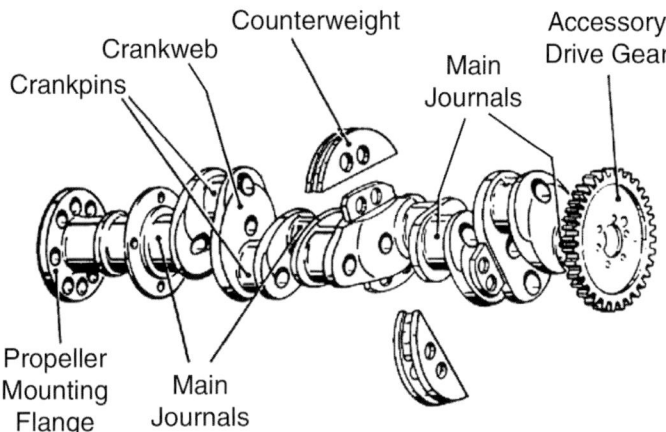

Fig. 1.11 A

The purpose of the crankshaft is to change the reciprocating motion of the piston and connecting rod into rotary motion for turning the propeller. Internal passages supply oil under pressure to all the bearings through oil-ways drilled in the main journals and crankpins.

A crankshaft consists of three main parts: a journal, a crankpin, and a crankweb. The number of **throws** classifies the crankshaft. A throw consists of two crankwebs and a crankpin. The length of the piston stroke equals the length of 2 crankwebs. There are as many crankthrows on a crankshaft as there are cylinders.

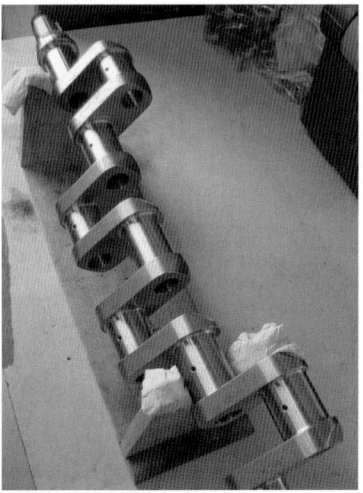

Fig. 1.11 B

The main bearing journals hold the crankshaft bearings, which in turn support the crankshaft. The bearings are usually plain, soft metal shell bearings and are easily replaced when worn. To dampen the torsional vibrations, counterweights are normally fitted to some of the crankwebs.

CONNECTING ROD

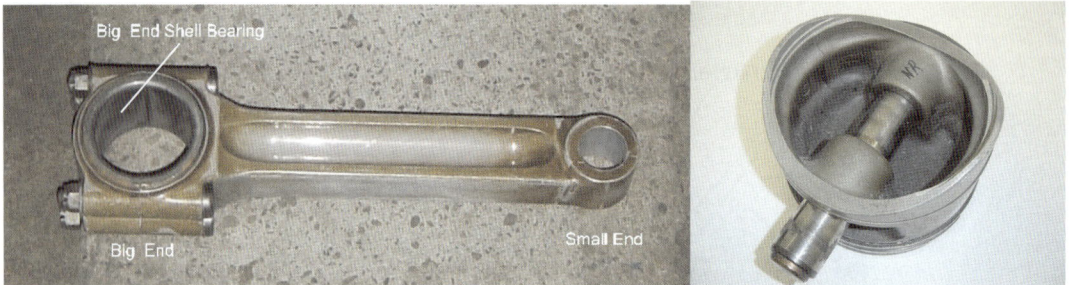

Fig. 1.12

The connecting rod links the piston to the crankshaft and transmits the force of the power stroke from the piston to the crankshaft. The connecting rod is attached to the piston by a free floating piston or gudgeon pin to distribute the wear around the pin and is referred to as the **small end**. The crankpin end is referred to as the **big end**. The big end bearings are similar to the main bearings with shell liners; whilst small end bearings may have a bronze insert.

PISTON

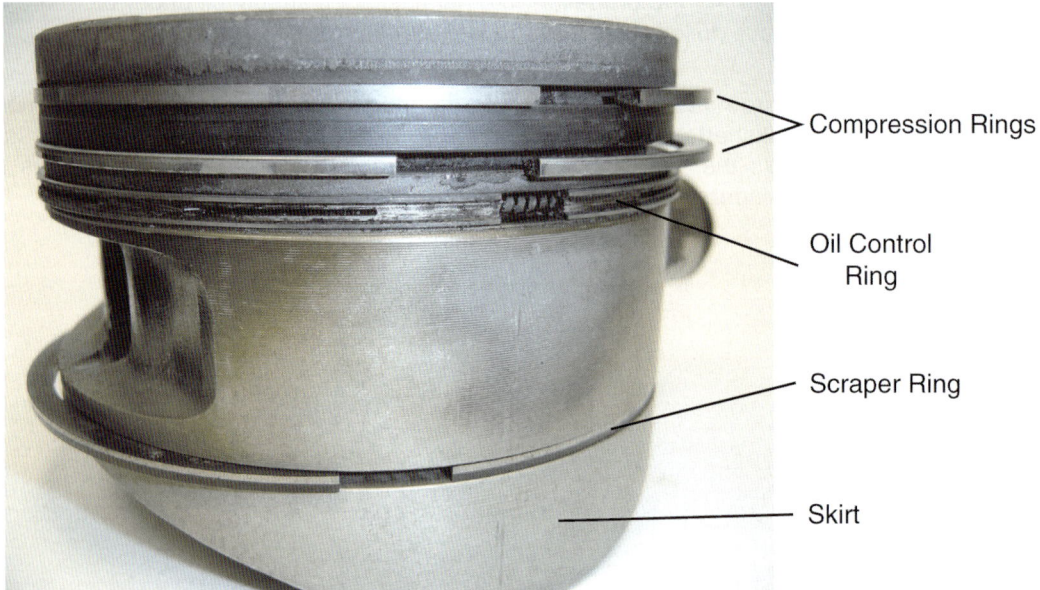

Fig. 1.13

The pistons are usually high-strength aluminium alloy forgings with the top of the piston being the crown. The sides around the bottom are called the skirt. They have grooves machined around them to hold the rings. The compression and oil control rings to form a sliding gas-tight plug in the cylinder. The rings nearest the piston crown are compression rings and prevent the gases in the combustion chamber from leaking into the crankcase. Oil control rings are installed in the lower grooves and regulate the thickness of the oil film on the cylinder wall. They are made of cast iron or alloy steel and have an expansion gap (figure 1.13).

CYLINDER BARREL

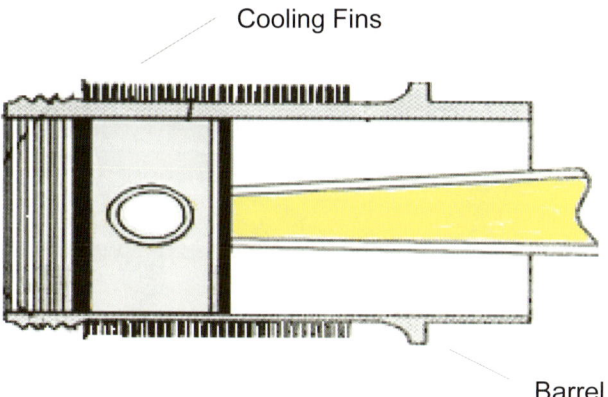

Cooling Fins

Barrel

Fig. 1.14

Alloy steel cylinder barrels provide a working surface for the piston rings. They must be strong enough to resist the pressure of combustion and must quickly dissipate heat. Cooling fins are machined on the outside, providing an increased surface area for cooling purposes.

CYLINDER HEAD

Fig. 1.15

For lightness and good heat dissipation, cylinder heads are usually made of aluminium alloy. They are usually screwed and shrunk on to the top of the cylinder barrels. Like the barrels, they have fins to aid cooling, see figure 1.15. The cylinder head provides a mounting for the rocker arm assemblies, valves, valve guides, seat inserts, springs, and sparking plugs.

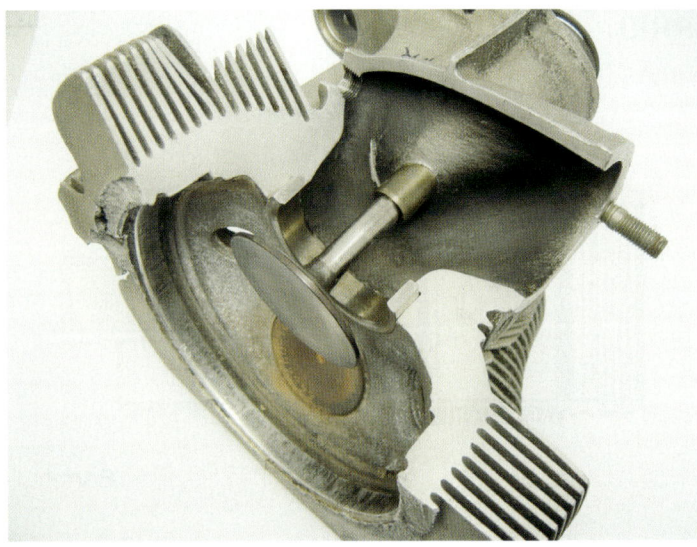

Fig. 1.16

The valves control the flow to and from the cylinder head via the intake and exhaust ports. Valve guides ensure the valves move in one line of motion only, therefore preventing rocking. They are usually pressed into the cylinder head, whilst valve seat inserts are ground to form a gas tight seal when the valves are closed. See figure 1.16.

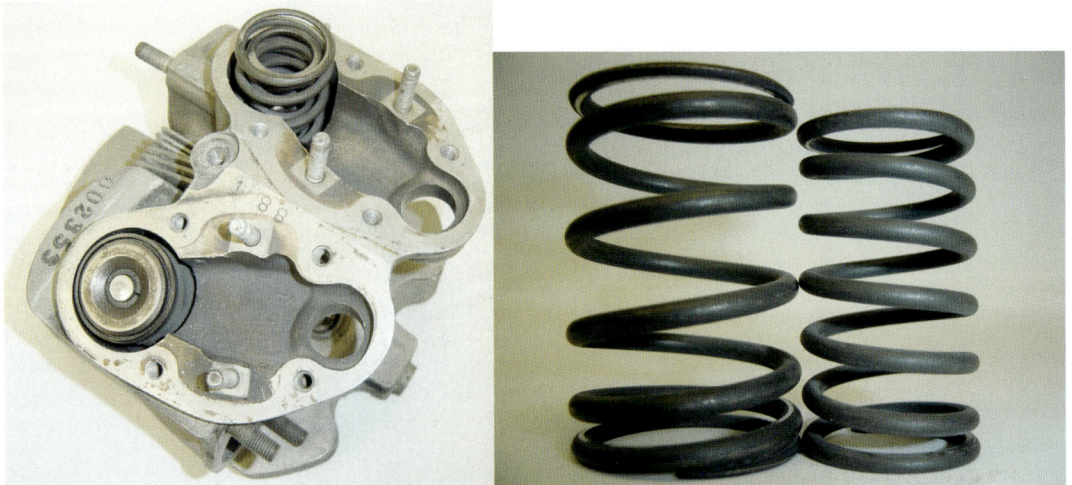

Fig. 1.17

Valve springs ensure that the valves remain closed except when opened by the rocker mechanism. Normally, there are two helical springs to each valve, coiled in opposite directions and of different thickness and diameter to help eliminate valve bounce and for safety reasons. Split collars hold them compressed between the cylinder head and the valve spring cap. See fig. 1.17. There are two threaded holes for spark plugs in each cylinder head.

VALVE OPERATION

A camshaft is driven at half engine speed, since the valves only operate once for every two revolutions of the engine. As the camshaft rotates, the high point of the cam bears on the tappet. This transmits vertical movement to a push rod that pushes up on the rocker arm causing it to bear down on the valve, opening it against the valve springs. Further rotation of the camshaft relaxes the operating mechanism and the springs close the valve. There is usually one cam lobe for each valve.

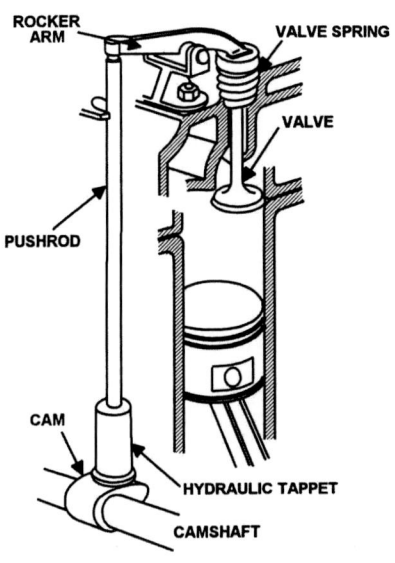

Fig. 1.18

A clearance must exist between the tappet and the pushrod to ensure that the valve closes completely. This tappet clearance causes noise and produces wear in the valve mechanism. Should the tappet clearance be out of adjustment, the valve timing is affected. If the gap is too large, the valves open late and close early, and vice versa. Fitting hydraulic tappets resolves this problem on most current engines. These operate by using engine oil as a hydraulic fluid and automatically adjust the tappet clearance, thus eliminating noise and wear, and reducing maintenance.

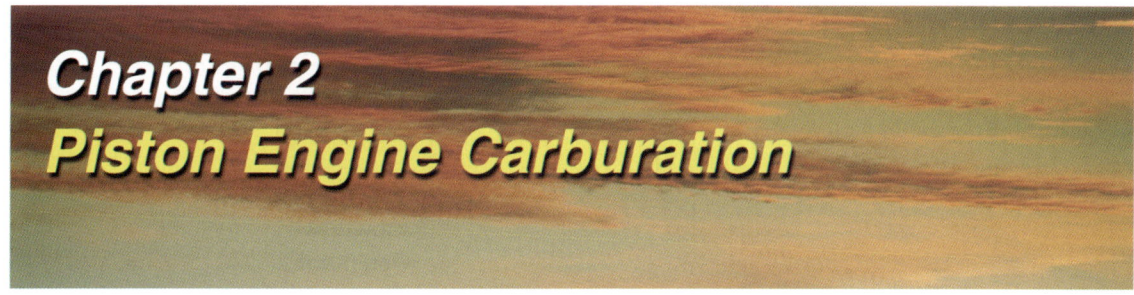

Chapter 2
Piston Engine Carburation

AVIATION FUELS

The fuel used for spark-ignition piston engines is a refined petroleum distillate comprised of one of the hydrocarbon families that consist of approximately 85% carbon and 15% hydrogen. When mixed with air and burnt, the carbon and hydrogen combine with the oxygen in the air to form carbon dioxide and water vapour.

AVGAS

Aviation gasoline is different from motor vehicle fuel, since it is subject to a more rigid control of quality assurance; and it has a much higher resistance to detonation. On piston-engine aircraft, it is important to use the correct type of fuel, since using the incorrect type of fuel can lead to low engine performance, detonation, and engine failure. There are at present three basic types of gasoline, which are dyed different colours for identification.

Grade 80 fuel has a low lead content, is only suitable for low compression engines, and is red in colour.

Grade 100 fuel has a high lead content, is used on high compression engines, and is green in colour.

Grade 100LL (Low Lead 100 Octane) fuel is a compromise between the Grade 80 and Grade 100, contains a medium lead content, and is blue in colour. This fuel is in general use.

Mogas (Automobile Fuel) fuel has a lower vapour pressure than AVGAS. Therefore, it tends to cause vapour locks in pipelines at high temperature and altitudes. Carburetted engines using this fuel are more susceptible to carburettor fuel icing. In addition, it has a low lead content, which can lead to detonation and pre-ignition. Before using MOGAS in an aircraft, consider all its disadvantages by first consulting Airworthiness Notice No. 98 and CAA General Aviation Safety Sense leaflet No. 4.

OCTANE RATING

This is a measure of the fuel's resistance to detonation; the higher the octane number, the higher its resistance. The aircraft flight manual or equivalent states the minimum octane rating. Never use a fuel with an octane number lower than that recommended. As already discussed, fuel octane ratings are colour coded (e.g. 100 LL BLUE).

FUEL CONTAMINATION

Fig. 2.1

Examine fuel on a regular basis for signs of contamination as listed below. Take a sample of fuel from the fuel drain points situated at the bottom of each fuel tank, fuel filter, and where applicable, cross feed lines.

> ➢ Globules of water
> ➢ More than a trace of sediment
> ➢ Cloudiness
> ➢ Positive reaction to water-finding paste, paper, or chemical detector

DENSITY OF FUELS

Specific gravity (SG), or relative density, is the mass per unit volume of a fuel and is compared with water at 15.5°C. When determining fuel loading, take variation of fuel density into account for the accuracy of the fuel contents and fuel flow. Temperature also has a marked effect on fuel density. As temperature increases, the density decreases. AVGAS typically has an SG of 0.72.

MIXTURE RATIO

During the combustion process, a chemical reaction takes place that requires a precise ratio of oxygen to gasoline. The weight ratio of air to gasoline that is required to ensure complete combustion of the fuel is 15-to-1, where 15 refers to air and 1 to fuel. The 15-to-1 ratio is the **chemically correct** or **Stoichiometric ratio** and is the theoretical ideal ratio. Mixture ratios vary between approximately 8-to-1 and 20-to-1 to cater for various engine requirements. Eight-to-1 is a **rich** mixture, and there is an excessive amount of fuel. Twenty-to-1 is a **weak** mixture, and there is an excessive amount of air. The **best power ratio** actually occurs at a richer mixture ratio of approximately 12-to-1 and is the mixture ratio that allows the engine to develop maximum power at a particular power setting. The Stoichiometric mixture (15:1) produces the highest combustion temperature. In rich mixtures, the excess fuel acts as a coolant (when changing from a liquid to a vapour, heat is extracted). In lean mixtures, less fuel is being burned (less heat), the burning rate is slower, and the same measure of air is better able to cool.

Due to imbalances that exist in the mixture ratio due to inefficient mixing and distribution, a variation of mixture strength can exist between cylinders. Slightly richer mixture strength is used, since the engine can function better on a slightly rich mixture than on a weak one. This is because a rich mixture has less of an effect on power than a weak mixture. At low engine speeds, some exhaust gases remain in the cylinder due to inefficient scavenging, resulting in the dilution of the mixture. Therefore, as engine speed decreases, the mixture should be enriched. A rich mixture is used for high power settings to use the excess fuel to aid cylinder cooling.

A weak mixture not only burns at a lower temperature than the chemically correct ratio, it also burns slower. Therefore, power output and fuel consumption both decrease. The weaker the mixture is, the greater the reduction in power. For range and economy, a weak mixture ratio is used. The **Best Economy Ratio** is the ratio that gives the lowest specific fuel consumption and occurs at approximately 16–18-to-1 as illustrated in figures 2.2 and 2.3.

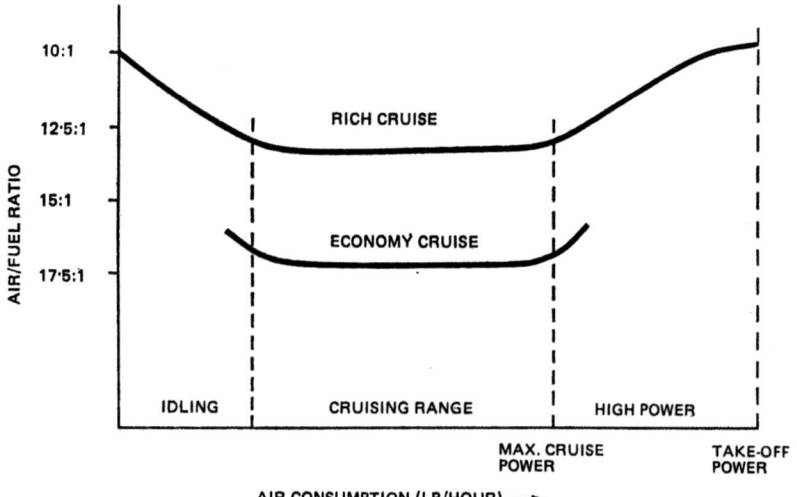

Fig. 2.2

Figure 2.2 illustrates typical mixture ratios that, depending on engine power, are used during engine operation. In this case, the maximum power occurs at a mixture ratio of 10-to-1, and the mixture ratio of 17.5-to-1 is used for maximum economy.

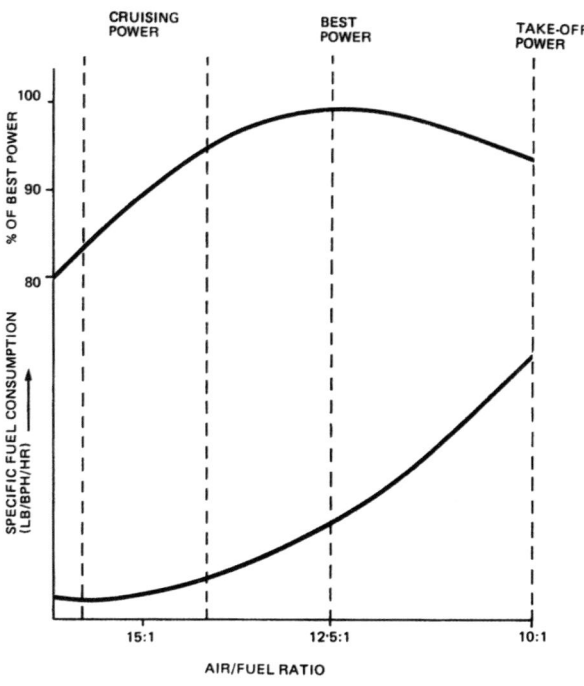

Fig. 2.3

Mixture ratios are, however, dependent on engine speed and power output. Figure 2.3 illustrates the specific fuel consumption and power relationship. The SFC reduces as the mixture strength weakens.

EXHAUST GAS TEMPERATURE

Use an EGT gauge to make accurate adjustments to the mixture ratio, since a rich mixture reduces the engine exhaust gas temperature, whilst a weak mixture increases it. Only carry out mixture adjustments using the exhaust gas temperature below 75% power and in the cruise.

The thermo-couple principle is used to measure EGT, which does not require power to operate. Two dissimilar metals are welded together at the ends. When heated, the voltage induced is proportional to the temperature difference between the two ends. The voltage is in millivolts and varies with the metals used. The hot end is fitted where the temperature is to be sensed; the other end, known as the cold end, is exposed to normal temperature.

FLAME RATE

This is the rate at which the flame front moves through the fuel and air mixture and is most rapid at the best power setting, falling off substantially either side of this setting. Too weak a mixture results in a slow flame rate. Therefore, the mixture is still burning when the inlet valve opens. This ignites the mixture in the inlet manifold and results in a backfire.

DETONATION

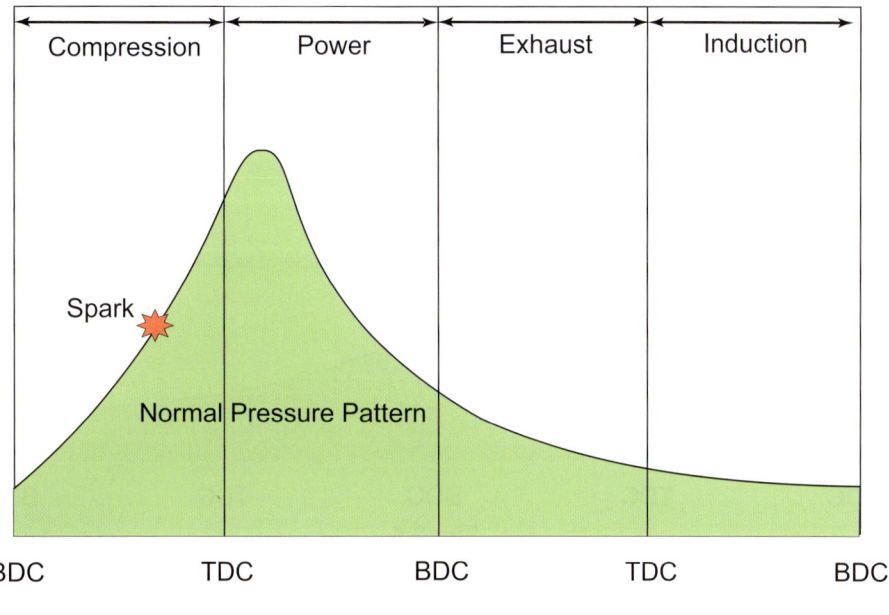

Fig. 2.4

The flame spread across the cylinder should be even and uniform. If the fuel is not sufficiently resistant to detonation, at a temperature and pressure critical to the fuel, spontaneous combustion occurs with a very high flame rate of approximately 1000 ft/sec as opposed to the normal flame rate of approximately 60-80 ft/sec. Since this occurs with such rapidity, there is an audible explosion referred to as **knock** or **pinking**, which is not normally heard inside the aircraft due to engine noise.

Fig. 2.5

When detonation occurs, the resulting excessive cylinder temperature causes a loss of power and possible engine damage. If allowed to continue, damage such as burning of the piston crown, valves, and valve seats, and seizing of the piston rings in piston grooves would result in engine failure. See figure 2.5, which shows a cylinder head that has split due to the excessive pressure. When adjusting power and/or mixture, follow laid down procedures in order to prevent exceeding engine limitations, especially the cylinder head temperature.

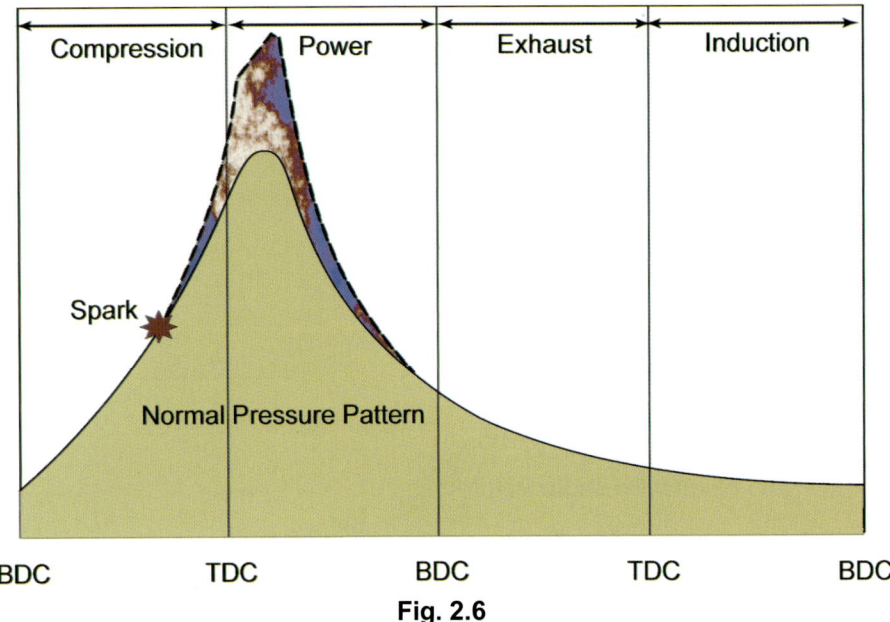

Fig. 2.6

Detonation takes place after spark ignition has occurred and affects the whole engine. Recognition is by a rise in cylinder head temperature and loss of power; rough running may also occur. If detonation is suspected, enrich the mixture and reduce power.

Increasing engine rpm reduces the effect of detonation, as the piston is retreating, increasing the volume, thereby reducing the pressure. Additionally, additives such as Tetraethyl Lead (TEL) are added to fuel to combat detonation.

Some likely causes of detonation are:

  ➤ Incorrect mixture strength
  ➤ Incorrect fuel grade
  ➤ Combinations of high pressure and temperature
  ➤ Time-expired fuel
  ➤ Ignition too far advanced

PRE-IGNITION

Pre-ignition affects individual cylinders and is due to a hot spot inside a cylinder or cylinders. The high temperature of detonation can cause the hot spots, which is some local area within the cylinder(s) that ignites the fuel/air mixture before spark ignition takes place. Operating the engine within its limitations and avoiding overheating can prevent pre-ignition. When pre-ignition is present, it is detected by a loss of power and rough running. Increasing the engine rpm makes the rough running worse. This is because the hot spot does not have time to cool between successive air-fuel charges at the higher rpm.

Chapter 3
Piston Engine Carburettors

INTRODUCTION

A carburation system must provide fuel and mix it with air in correct proportions under all engine operating conditions irrespective of altitude and attitude. A carburettor, pressure carburettor, or direct fuel injection system delivers the mixture evenly to the cylinders in the correct state for efficient combustion.

SIMPLE FLOAT CARBURETTOR

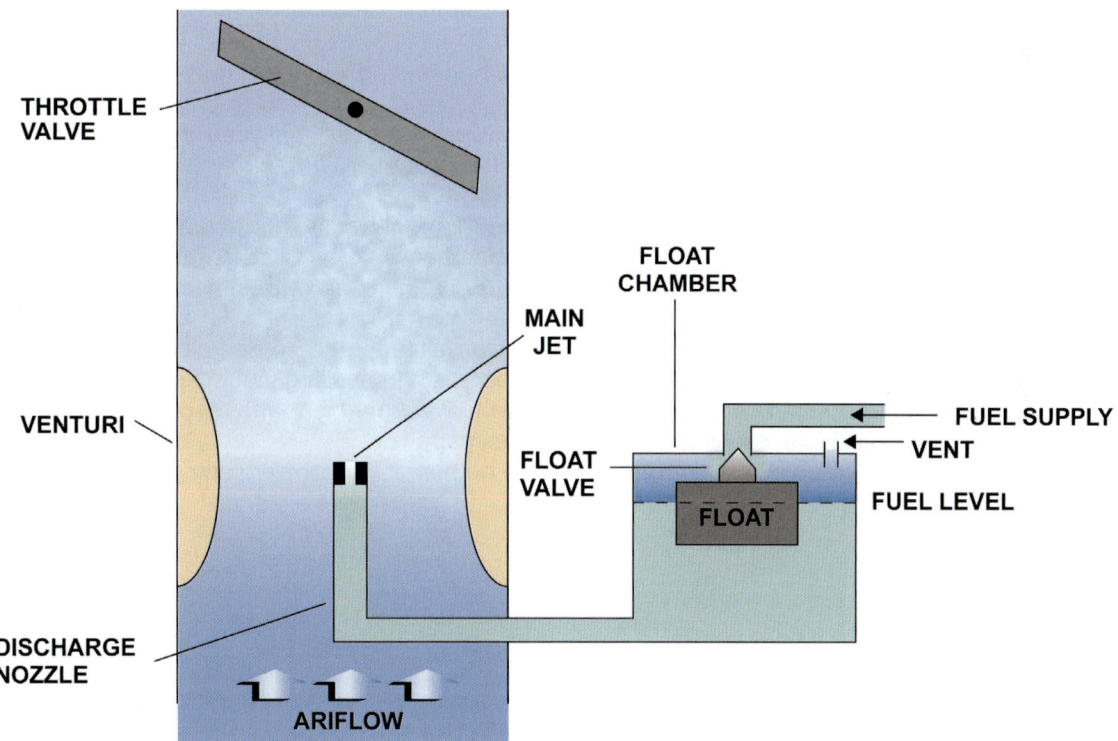

Fig. 3.1

The constituent parts of a simple float carburettor are the float chamber, main jet, venturi (also referred to as the choke), and throttle valve.

The suction created during the induction stroke draws air from the air intake, through the carburettor venturi, and on to the inlet manifold to the cylinder. As the air flows through the venturi, it accelerates causing a reduction in static pressure. This creates a pressure difference between the float chamber and the venturi. The aircraft fuel tanks supply fuel to the float chamber via gravity, an electric boost pump, or engine-driven pump. The float needle valve maintains a constant level.

The float chamber is subjected to air intake pressure or atmospheric pressure that acts on the fuel. The float chamber connects to a fuel discharge nozzle that is located in the venturi throat. The pressure differential that exists between the float chamber and the venturi causes the fuel to flow from the float chamber to the discharge nozzle and into the airstream via the main jet. Opening the throttle increases the airflow through the venturi. This results in a greater pressure drop at the venturi throat, which, in turn, increases the pressure differential causing more fuel to flow. This causes the fuel level in the float chamber to fall, allowing the float to lower and open the float valve to admit more fuel into the float chamber.

At steady running conditions, the float takes up a sensitive position, matching inflow to outflow for that condition, and the fuel level in the float chamber remains constant. Conversely, closing the throttle reduces the pressure differential, fuel flow decreases, and the float rises. The float needle valve reduces the amount of fuel flowing into the float chamber until it reaches its sensitive position once more. Fig. 3.1 illustrates a simple float carburettor.

LIMITATIONS

The simple float carburettor only provides a suitable air/fuel mixture over a very limited range. Therefore, a number of additions to the simple float carburettor are necessary for aircraft engine requirements, such as changes in engine speed, aircraft forward speed, aircraft altitude, and power requirements. These additions are necessary to overcome the following limitations:

- ➤ The mixture progressively becomes rich when increasing rpm due to the different responses of fuel and air.
- ➤ At low engine speed, a very low differential pressure exists between the float chamber and the venturi. As a result, enough suction does not exist to draw fuel from the float chamber.
- ➤ Since the float chamber is open to the atmosphere, it results in uneven fuel flow due to small air pressure changes within the float chamber. As a result, the source of air to the float chamber must be at constant pressure or increase with mass flow within the induction system.
- ➤ Whilst maintaining constant rpm with increasing altitude, the same volume of air enters the cylinders. However, the weight of air is decreasing, resulting in a richer mixture.
- ➤ Rapid engine acceleration causes the mixture to weaken due to the response differences between air and fuel, causing a temporary loss of power.
- ➤ To prevent overheating, the mixture needs to be rich at high power settings.

DIFFUSER

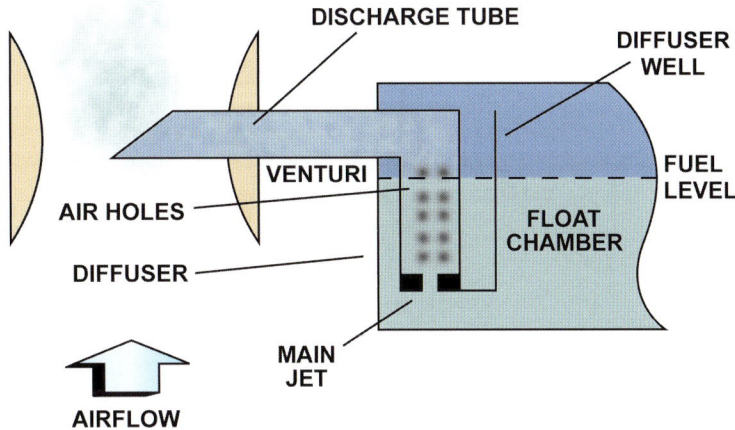

Fig. 3.2

A diffuser is a device to prevent the mixture from becoming rich with increasing rpm, as illustrated in figure 3.2. When the engine is at rest, the fuel level in the diffuser is the same as the float chamber. When the engine is idling, the fuel level in the diffuser significantly decreases and uncovers some of the air holes, allowing air into the discharge tube.

As the engine rpm increases above idle, the fuel level within the diffuser drops progressively, uncovering an increasing number of air holes. More air enters the discharge tube, thereby decreasing the pressure differential and preventing mixture enrichment. As a result, the fuel is vaporised more readily, especially at low engine rpm, thus aiding the combustion process.

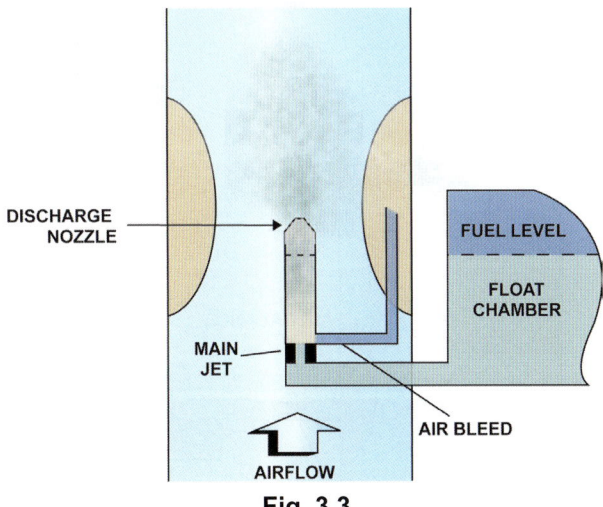

Fig. 3.3

Alternatively, installing an air bleed system, as shown in figure 3.3, offers the same function as the diffuser. Where air at atmospheric pressure is bled into the fuel discharge nozzle via the air bleed line, the resulting additional air opposes the enriching effect, maintaining a constant air/fuel ratio.

IDLE OR SLOW RUNNING SYSTEM

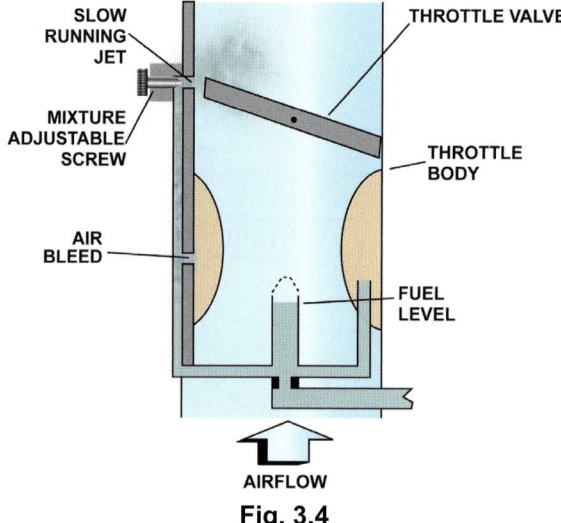

Fig. 3.4

At idle or slow running, the engine speed does not pull sufficient airflow through the venturi to provide a satisfactory discharge of fuel. Fuel for slow running enters the inlet manifold via a slow running jet located near the edge of the throttle valve where the static pressure is low (figure 3.4). This is due to the acceleration of air through the gap between the wall of the induction system and the throttle valve. The float chamber is linked to the discharge ports and slow running jet. The air bleed provides a mixture of air and fuel to the slow running jet.

The mixture is enriched to overcome the effects of mixing exhaust gas with fresh mixture during valve overlap. As the engine rpm is increased above the idle or slow running values, the system no longer operates, and fuel is now discharged in the normal manner.

AIR PRESSURE BALANCE

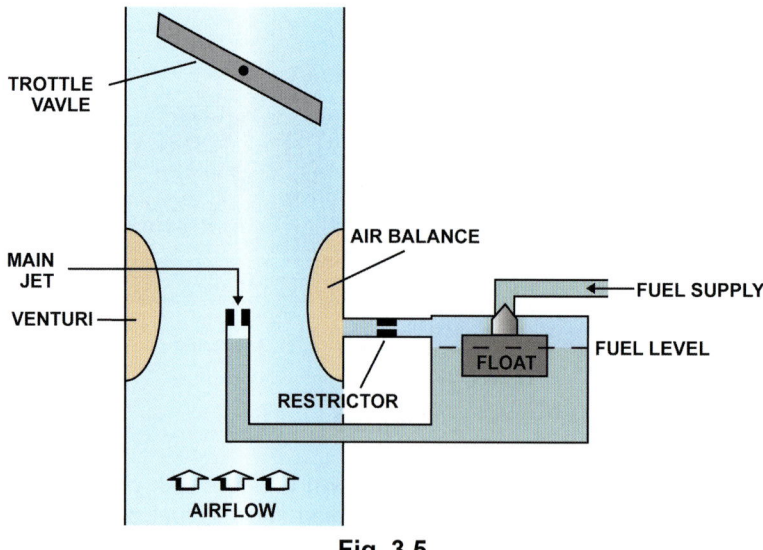

Fig. 3.5

Float chamber air pressure is essential for the float carburettor to operate satisfactorily. To have the float chamber vented to atmosphere is unsatisfactory due to varying static pressure surrounding the carburettor. Therefore, air to the float chamber is normally taken from the rear of the venturi, and the leading edge of the venturi is open to receive impact airflow pressure. A restrictor is fitted in the air pressure balance line in an effort to maintain a constant or, with increased speed, a slightly increased float chamber pressure (figure 3.5).

ACCELERATOR PUMP

If the throttle valve opens rapidly, the airflow responds almost instantaneously, and a larger volume of air flows through the carburettor. The fuel flow, however, responds slowly to the changing conditions and a temporary weakening of the mixture occurs, causing a flat spot (loss of power). An accelerator pump injecting fuel into the induction system overcomes this weakening effect when the throttle opens quickly. A direct link between the accelerator pump and the throttle forces fuel into the venturi whenever the throttle opens. Some accelerator pumps allow the throttle to open slowly without discharging fuel into the inlet manifold by a controlled bleed of fuel past the pump piston. Other accelerator pumps incorporate a delayed action plunger. Figure 3.6 shows this type of pump.

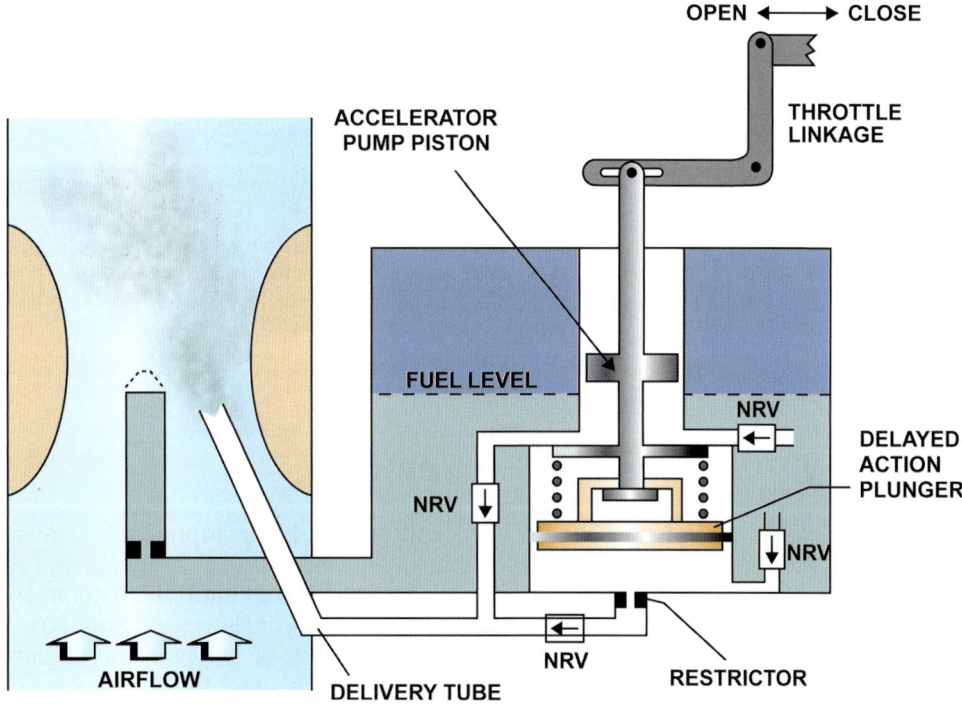

Fig. 3.6

The system illustrated consists of a cylinder and a piston that connects to the throttle valve linkage. As the throttle is closed, the piston moves up, thus charging the cylinder with fuel from the float chamber via a non-return valve (NRV). As the throttle opens, the accelerator pump piston moves down, closing the inlet non-return valve, forcing the fuel out past the discharge non-return valve, through the accelerator delivery tube and into the airflow.

Some accelerator pumps have a delayed action plunger to continue the fuel flow for a few seconds after throttle movement has ceased. By fitting a restrictor below the delayed action plunger, a spring compresses as the accelerator pump piston moves down. Once the initial acceleration has taken place, the spring expands, forcing down the delayed action piston and a subsequent fuel flow through the restrictor into the delivery tube and airflow.

MIXTURE CONTROL

For accurate adjustment of the mixture ratio both for cruise power settings and for high altitudes, a manual mixture control is required. During the climb, the air density decreases, and the mixture progressively enriches. Fuel flow must be reduced to maintain the correct ratio. Some large engine carburettors have an automatic mixture control for altitude. There are two basic types of manual mixture control, which are described below.

NEEDLE TYPE MIXTURE CONTROL

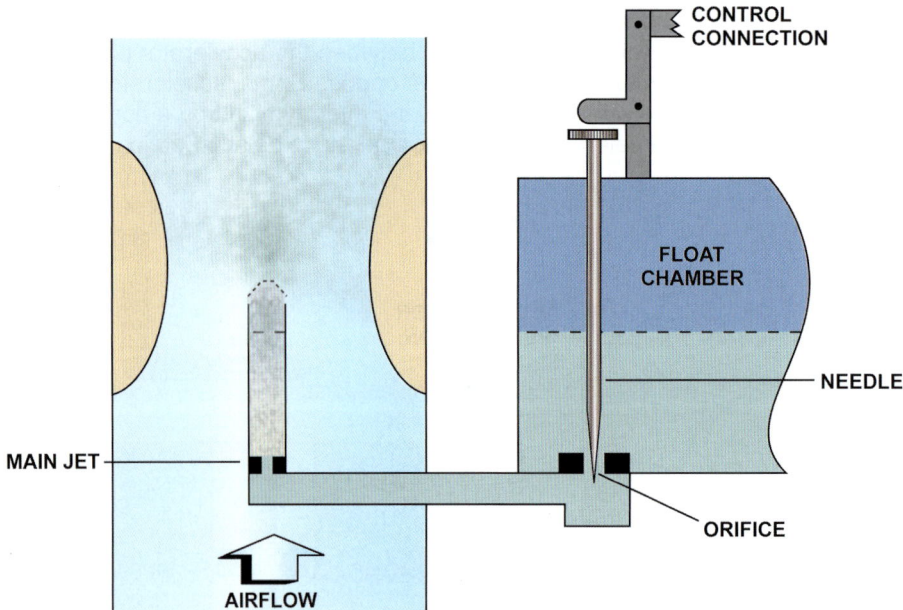

Fig. 3.7

This type of mixture control consists of a cockpit control lever connected to a needle valve situated in the float chamber. Movement of the lever either raises or lowers the needle, thus varying fuel flow through an orifice to the main jet. The position of the needle controls the mixture strength. Moving the lever fully rearward positions the needle fully down. Fuel flow through the orifice to the main jet is stopped, thereby providing a means of stopping the engine. When the lever is in this position, it is in the **idle cut-off** position and is used for engine shut down. When the lever is fully forward, the needle is fully up allowing the maximum fuel flow to the main jet, which results in a fully rich mixture. Intermediate positions of the lever adjust mixture ratio for all operating requirements.

AIR BLEED MIXTURE CONTROL

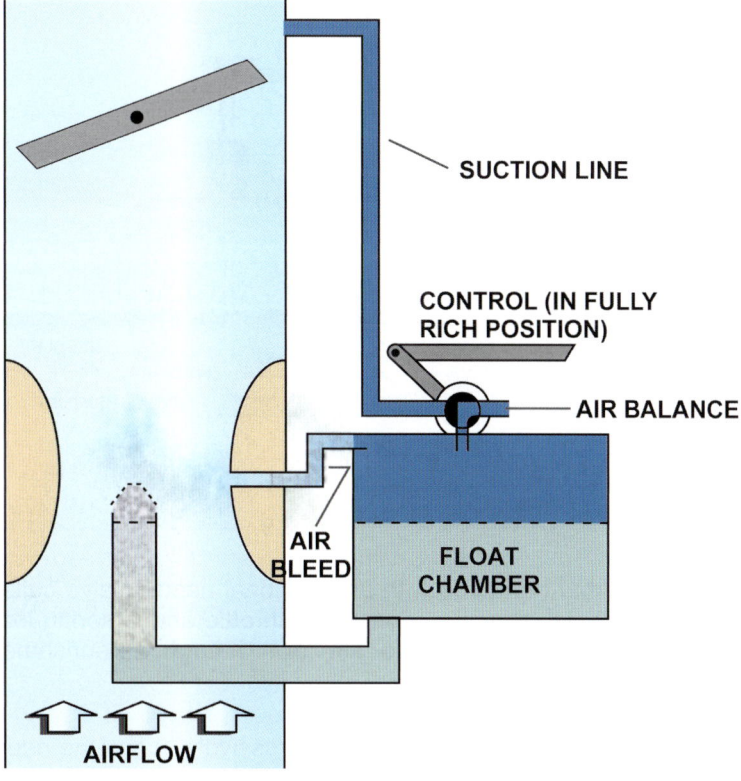

Fig. 3.8

This system functions by controlling the air pressure in the float chamber and varying the differential pressure that the fuel is subjected to. Between the venturi and the float chamber is a small air bleed that tends to reduce the air pressure in the float chamber. A control lever in the cockpit is connected to a valve, which controls airflow into the float chamber.

When the lever is fully forward, the valve is fully open, the air pressure is at its highest, and the mixture is fully rich. Moving the lever rearward progressively closes the valve and air pressure reduces, therefore fuel flow reduces, weakening the mixture. In figure 3.8, a suction line to the engine side of the throttle valve is incorporated. When the control lever moves fully rearward to the **Idle Cut-Off** position, the suction line is connected to the float chamber. Due to a lower pressure existing downstream of the throttle, air pressure in the float chamber is reduced and the fuel flow ceases, stopping the engine.

POWER ENRICHMENT/ECONOMISER SYSTEMS

In order to prevent detonation at power settings above the cruising range, a rich mixture is required. At full throttle, an additional fuel supply that absorbs heat as it changes into vapour may provide a rich mixture. Alternatively, an economiser system may be installed that provides a rich mixture at high power and bleeds off float chamber air pressure to allow the engine to operate with a relatively weak mixture for all conditions except full power. One of the two methods normally achieves power enrichment.

NEEDLE VALVE ENRICHMENT

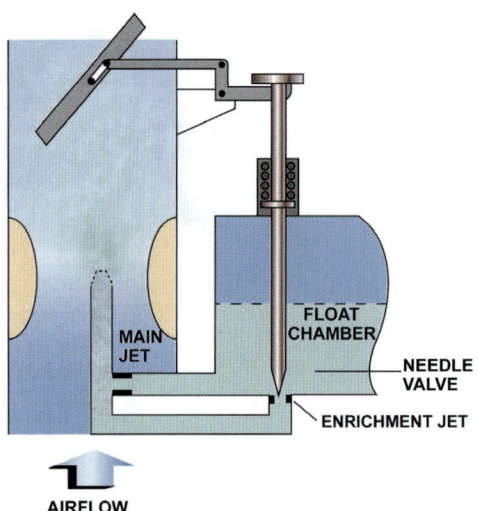

Fig. 3.9

Figure 3.9 shows a carburettor fitted with an additional needle valve called either a power jet or enrichment jet. The needle valve connects to the throttle and is spring-loaded fully closed at the maximum sea level cruising condition. No fuel flows through the enrichment jet when it is closed and the main jet does the metering.

Opening the throttle above the cruise setting progressively opens the needle valve, compressing the spring and admitting more fuel to the discharge tube until it is fully open at full throttle with maximum fuel flow through the enrichment jet. On closing the throttle, the needle valve is progressively closed by the action of the spring, and the fuel flow through the enrichment jet reduces until it is fully closed at the cruise setting.

BACK-SUCTION ECONOMISER

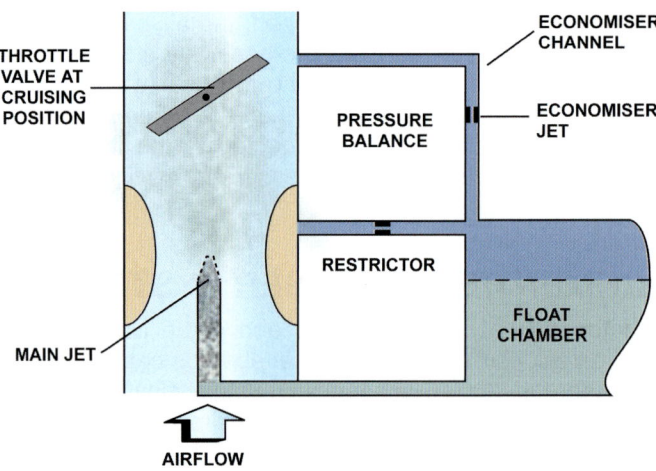

Fig. 3.10

At high power settings, air flowing past the throttle valve is almost at atmospheric pressure, which has a minimum influence on the float chamber pressure, therefore providing a rich mixture at this setting. Closing the throttle to the cruise condition creates suction as the air flows between the throttle valve and the wall of the inlet manifold. This causes air to flow from the float chamber through the economiser jet, thus reducing the pressure in the float chamber and subsequently reducing the fuel flow through the main jet to provide an economical mixture for the cruise.

CARBURETTOR AND INTAKE ICING

Awareness of icing in the carburettor and intake is very important for safe operation. Knowledge of the current Aeronautical Information Circular on induction system icing is of importance.

When liquid evaporates, it absorbs heat from its surroundings. Therefore, when fuel is sprayed into the induction system, it extracts heat from its surroundings. This causes the water vapour to condense, and it freezes if the temperature is low enough. The following six evaporation factors have an influence on the rate of formation of ice in carburettors:

> ➤ Temperature increases
> ➤ Surface area of the liquid increases
> ➤ Atmospheric pressure decreases
> ➤ Humidity of the air decreases
> ➤ A direct function of the volatility of the exposed liquid
> ➤ Air flow across the surface of the liquid increases

CARBURETTOR ICE FORMATION

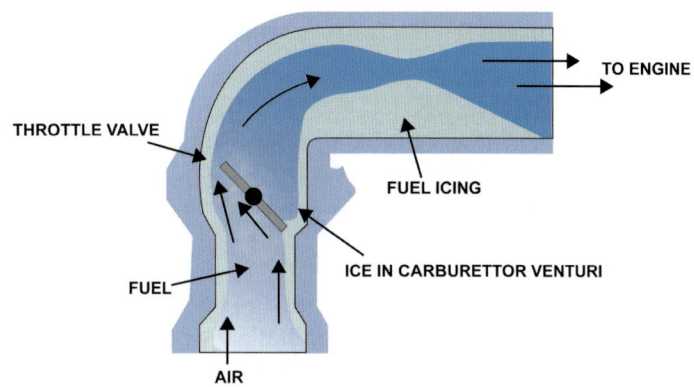

Fig. 3.11

As fuel is sprayed into the low-pressure area of the carburettor venturi, it rapidly evaporates, cooling the air, the wall of the induction system, and the water vapour. If the air humidity is high and the metal of the carburettor is cooled to below 0°C, ice forms and operation of the engine is affected. This is due to the size and shape of passages, which a coating of ice changes.

Carburettor ice formation may occur by any one of three processes:

> ➤ The freezing of the condensed water vapour of the air at or near the throttle forms ice, which is known as **throttle ice** or **expansion ice**, and is the most likely form of icing.
> ➤ The cooling effect of the evaporation of the fuel after it is introduced into the airstream may produce what is known as **fuel ice** or **fuel evaporation ice.**
> ➤ Water in suspension in the atmosphere coming into contact with engine parts at a temperature below 0°C may produce **impact ice** or **atmospheric ice.**

THROTTLE ICE

With the throttle in a partially closed position, such as when descending, throttle icing is most likely. In this position, the air velocity at the edge of the throttle valve is increased, and a pressure and temperature drop occurs, causing ice formation on the throttle plate. It is important to remove throttle ice as quickly as possible. Due to ice build up on the throttle, the venturi effect between the throttle and the wall of the inlet manifold increases, causing an even greater pressure and temperature drop, therefore making the situation worse.

IMPACT ICE

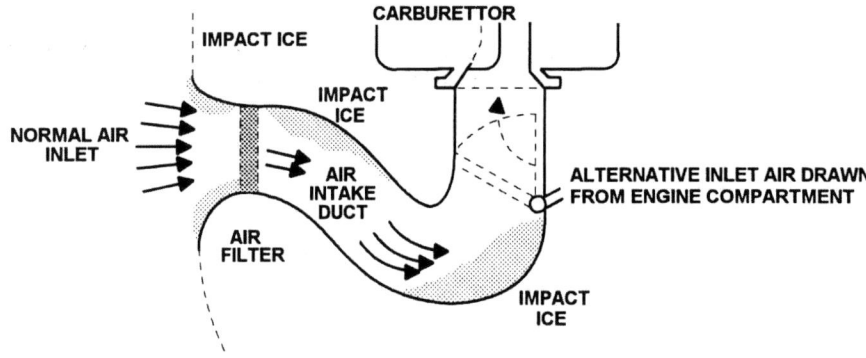

Fig. 3.12

Water droplets freeze when they impact with the intake and throttle body wall, which is most likely to occur at temperatures of between 0°C and -7°C. The ice builds up around the air filter and intake, choking the airflow, and alters the fuel/air ratio, causing a loss of power or engine failure.

CARBURETTOR HOT AIR CHECK

Carry out this check immediately before take-off. Note that if carrying out this check some distance from the runway requiring extensive taxiing, ice can re-form in the induction system. Follow the procedure in the flight manual or other approved document. Generally, the procedure is as follows:

> ➢ Select the required rpm and select HOT on the hot air lever, noting the rpm decrease.
> ➢ If there is no decrease apparent in rpm, then hot air is not entering the induction system.
> ➢ Should rough running occur, this indicates ice formation in the induction system; therefore, continue in the hot selection until rough running ceases. Then select COLD, noting the increase in rpm. Decreased rpm is due to the reduction of air density of the heated air.
> ➢ Take-off should not be attempted with hot air selected, since power is reduced and the likelihood of detonation is increased.

CARBURETTOR INTAKE HEATING

In order to remove ice or to prevent ice formation during flight induction, heating must be provided. Intake heating is achieved by supplying hot air from a jacket around the exhaust system. Cold air heated by the hot exhaust flows into the induction system. An adjustable valve operated by a heat control lever located in the cockpit controls the hot air entering the system. Normally, the selection is either fully hot or fully cold with no intermediate position. When hot air is selected, it is not filtered, and abrasive material can enter the cylinders, increasing the rate of wear of the cylinder barrels. Furthermore, hot air reduces density and hence power; this also increases the risk of detonation.

Some installations route the induction system manifold downstream of the carburettor through the oil sump to pre-heat the pipe work. This helps in vaporisation and reduces the risk of ice formation. Injection systems require very little or no hot air supply. However, they do require an alternate air supply to bypass the air filter should it become blocked.

EFFECT OF INDUCTION SYSTEM ICING ON ENGINE PERFORMANCE

Recognition of carburettor icing and the weather conditions associated with icing is very important. One or more of the following can indicate carburettor icing:

> ➢ Lower than normal cylinder head temperature
> ➢ Rough running engine
> ➢ Loss of power
> ➢ Loss of altitude
> ➢ Loss of airspeed

The enriching effect causes cylinder head temperature reduction and is followed by rough running and power loss or even a complete power loss in the most serious cases. A drop in engine speed indicates this power loss with fixed pitch propellers, whilst with constant speed propellers, a drop in manifold pressure, with the rpm remaining constant, indicates a reduction in power.

The normal procedures to prevent icing are:

> ➢ Check the carburettor hot air system before take-off
> ➢ Select carburettor heat during descent
> ➢ Select heat whenever icing conditions are suspected

It is a misconception that carburettor icing does not take place when the ambient air temperature is above 0°C (see figure 3.13). In fact, it can occur at temperatures as high as 30°C under the right conditions. It can occur under combinations of humidity and temperature as indicated in the graph below. Should ice be present and carburettor hot air selected, engine roughness is initially likely to get worse before it disappears. This is due to the dramatic cooling effect of water or water vapour flooding into the cylinder. Strictly adhere to the instructions concerning the use of carburettor heat or alternate air control that appear in the aircraft's flight manual or equivalent document.

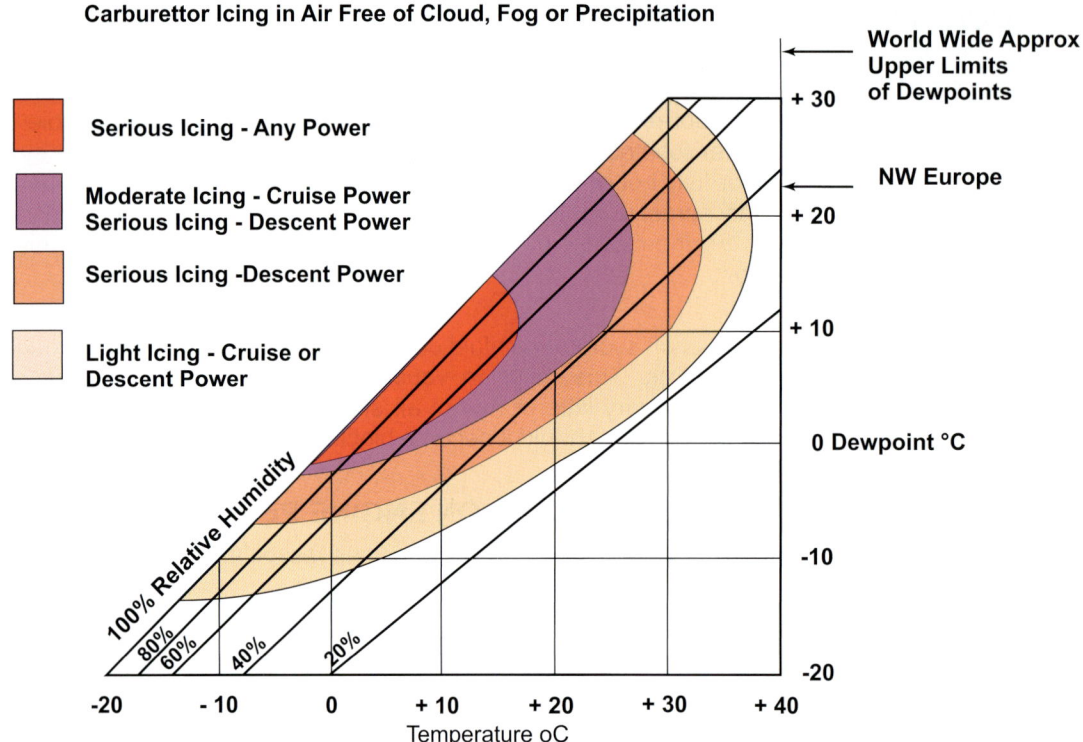

Risk and Rate of Icing Will Be Greater When Operating In Cloud, Fog and Precipitation

Fig. 3.13

Chapter 4
Piston Engine Lubrication and Cooling

INTRODUCTION

The primary purpose of lubrication is to reduce friction created between the moving parts of the engine. Good lubrication can substantially reduce engine wear, therefore prolonging engine life. Placing a film of oil between the moving parts reduces friction, preventing contact of the metal surfaces. Therefore, movement is between the layers of oil and not the metal surfaces.

Oil also seals between moving parts. For example, applying an oil film on the cylinder walls and piston forms a seal in the cylinder, which prevents gas leaks from the combustion chamber. In addition, oil protects against shocks between engine components such as the crankshaft, connecting rods, and valve operating mechanism by applying a film of oil that cushions the shocks.

Another important function of the oil as it circulates around the engine is to absorb heat from the internal engine components. This heat dissipates to atmosphere by passing the oil through an air-cooled oil cooler (heat exchanger). As the oil circulates, it collects contamination in the form of dirt, dust, and carbon that the atmosphere and the combustion process have introduced into the engine. Finally, the oil provides protection from corrosion of the internal metal parts.

LUBRICATING OIL TYPES

There are various types of oils available to fulfill the requirements of engine operation.

Straight mineral oil, or straight oil, is normally used after maintenance or when running in a new engine, where it is used for the first 50 hours of engine life. This type of oil can cause sludge to form that may result in clogging of oil ways and filters. Engines that do not require ashless dispersant oil can use straight mineral oil.

Ashless dispersant oil contains a dispersant that holds contamination in suspension, therefore preventing the formation of sludge that can occur with straight mineral oil. Contaminants are deposited safely in the filter rather than the engine. This oil cannot be mixed with straight mineral oil. As a result, it is essential to ascertain what type of oil an engine is using.

Synthetic oil is superior to the other oils mentioned above in all aspects, but due to expense and limited service experience, few piston engine manufacturers approve it. It is usually required, however, in gas-turbine engines.

OIL GRADES

Oil grades are determined according to their **viscosity**, where viscosity is defined as the fluid friction or the resistance to flow. This is very important in engine operation. High viscosity oil is thick and, therefore flows slowly, whilst low viscosity oil is thin and flows freely. Oil is required to maintain viscosity in order to withstand high bearing pressures and temperatures. Changes in temperature affect the viscosity of oil. Temperature increases thin the oil, allowing it to flow more freely (i.e. lower its viscosity and vice versa). Therefore, at high ambient temperatures, high viscosity oil is used and low viscosity oil is used at low ambient temperatures. The oil selected for use depends on the average ambient temperature. It is essential to use the correct oil grade for efficient lubrication.

Oils are graded by numbers that indicate their viscosity; the higher the number, the higher the viscosity and the slower the oil flows and vice versa. The numbers are obtained by using the Saybolt Universal Viscometer, which times a measured amount of oil at a particular temperature as it flows through a calibrated orifice. If it takes 20 seconds to flow through the viscometer, it receives the grade SAE 20, where SAE stands for the Society of Automotive Engineers. For commercial aviation, a number double that of the SAE number (i.e. SAE 20 equals commercial aviation grade 40) identifies the grades.

The letter W is used when grading oil. When the W is after the number, this indicates that the oil is satisfactory oil for winter use (e.g. 40W). Alternatively, a W before the number indicates that it is ashless dispersant oil (e.g. W80).

MULTI-GRADE OILS

Oils are now produced to meet the requirement of more than one grade. For example, multi-grade oil SAE 20W/40 possesses a viscosity within both the SAE 20W and SAE 40 range from 18°C to 99°C, giving it near constant viscosity and a wider performance range than single-grade oils. The dotted line on the graph indicates this. Although multi-grade oils are approved for aero-engines, they are not generally used.

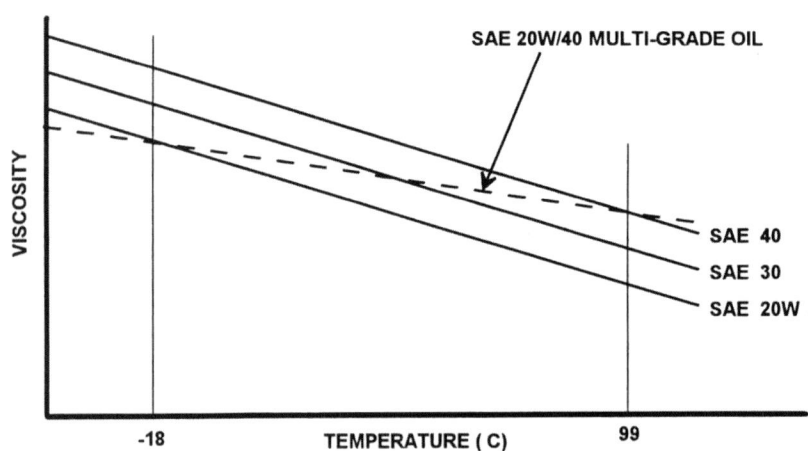

Fig. 4.1

LUBRICATION SYSTEMS

There are three methods used to supply oil for lubrication of engine parts: splash, pressure, or a combination of splash and pressure. A splash system relies on crankshaft rotation, gears, or special flingers to distribute the oil in a heavy mist form. The oil is splashed inside the engine crankcase by the gears or flingers dipping into the oil in the sump, lubricating pistons, cylinder walls, piston pins, some of the timing gears, and in some installations, the camshaft bearings. The splash system is never used on its own in aircraft engines.

In the pressure system, a pump takes oil from a tank or sump and delivers it under pressure directly to the internal components. This ensures a positive supply of oil to areas such as crankshaft bearings, camshaft bearings, and hydraulic tappets, and in the case of a constant speed propeller, the propeller governor. Depending on the system, the return oil is either returned to a tank by means of a scavenge pump or by gravity to the sump at the bottom of the engine. Aircraft engines use either the pressure system, or a system combining splash and pressure.

There are two basic types of lubrication system, dry sump and wet sump.

DRY SUMP SYSTEM

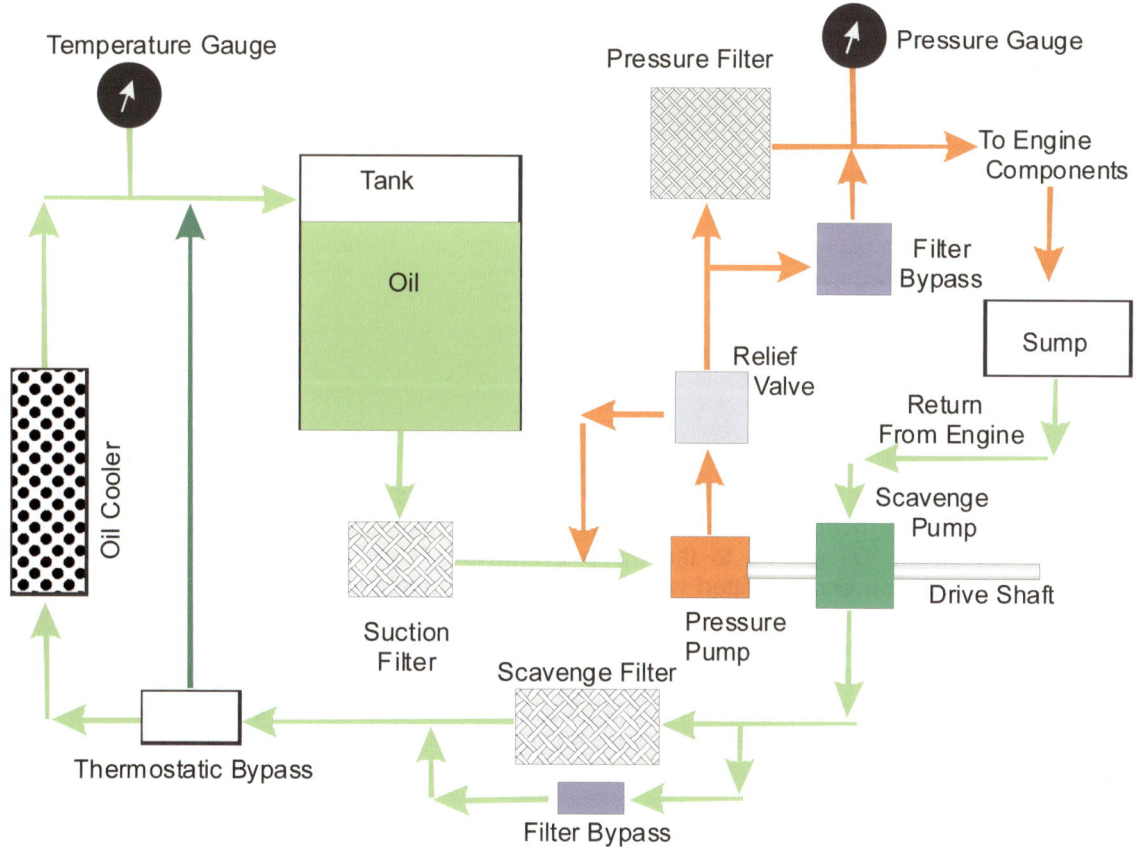

Fig. 4.2

With this system, the oil is stored in an externally mounted oil tank and is delivered to the engine by means of a pressure pump, via a pressure-relief valve and a pressure filter. After lubricating the engine, the oil is collected in a sump from where it is returned to the tank by means of the scavenge pump via a scavenge filter.

Dependent upon oil temperature, a thermostatic valve routes the return oil either through the oil cooler or directly back to the tank. A pressure-relief valve is located downstream of the pressure pump, determining and controlling the maximum system working pressure. The scavenge pump has a greater capacity than the pressure pump and prevents oil from accumulating in the sump.

WET SUMP SYSTEM

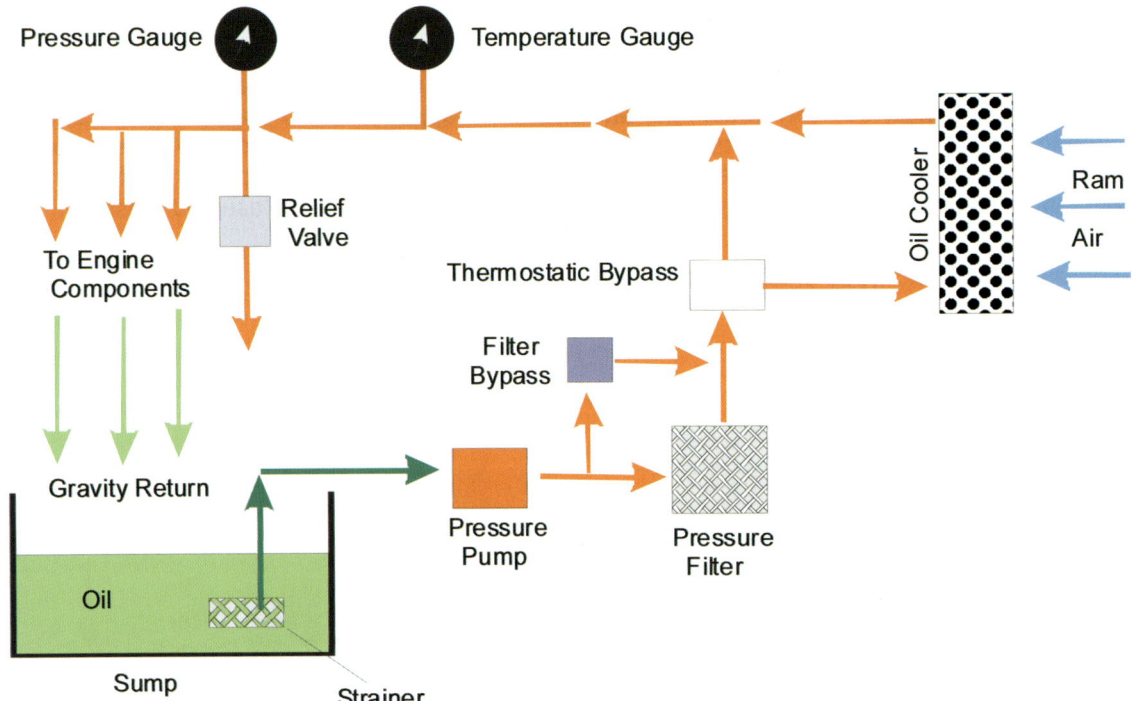

Fig. 4.3

The wet sump system is the type now used on modern light aircraft, and the oil is contained in the sump located on the bottom of the crankcase. This system normally incorporates splash and pressure lubrication. Oil returns to the sump by gravity, therefore a scavenge pump is not required. However, on engines fitted with a turbocharger, a scavenge pump is usually fitted to scavenge the turbocharger bearing that is lubricated by the engine oil system. In the schematic system shown below, the pump draws the oil via a strainer from the sump. The oil then flows to a pressure filter, which is fitted with a bypass in case of a filter blockage, then onto the oil cooler via a thermostatic bypass and on to a spring-loaded pressure-relief valve. Oil then enters the engine and flows through drilled passages to provide lubrication for all the engine internal components. For an engine fitted with a variable pitch propeller, oil is also supplied for operation of the propeller governor.

SYSTEM COMPONENTS

PRESSURE PUMP

The pump is normally a gear type pump driven by the engine. Its purpose is to deliver the oil to the engine at the correct pressure and ensure adequate lubrication. It consists of two gears, one of which is driven by the engine (termed the driven gear), which meshes with and drives the other gear. The two rotating gears capture oil in the spaces between the gear teeth and the pump housing on the suction side, transferring it to the pressure side.

The pressure obtained depends on the downstream flow restrictions created by bearings, etc. The pressure pump must be able to deliver sufficient oil to ensure adequate lubrication over the complete engine speed range. It is a positive displacement pump, where the same amount of oil flows through the pump at a consistent rpm. To ensure that the system pressure remains constant because of engine speed changes and, therefore, pump speed variations, there must be some method of relieving excess pressure.

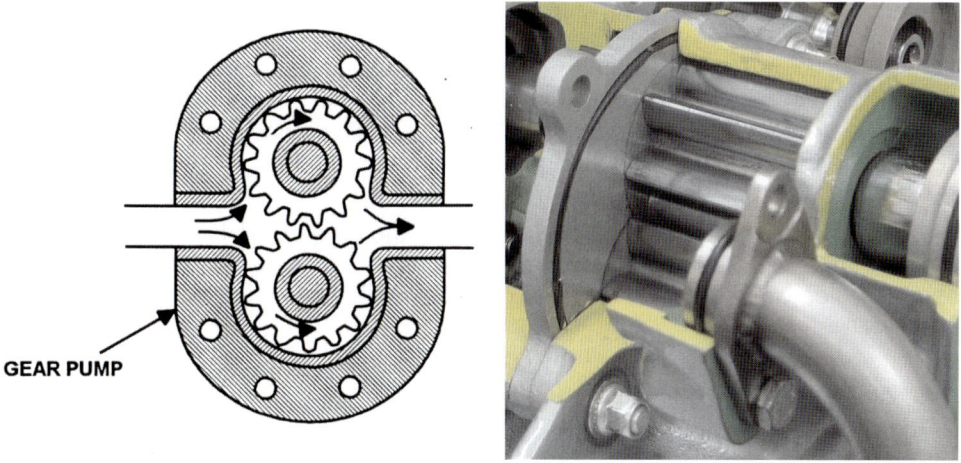

Fig. 4.4

Fitting a pressure-relief valve achieves this. If the pressure is below that for which the relief valve is set, the valve remains closed. When the pressure exceeds the pre-determined value, the valve opens returning oil to the sump or inlet side of the pump. This ensures that the system pressure is constant. The relief valve can be a simple spring-loaded ball arrangement or, for high altitude, a more complicated arrangement that senses atmospheric pressure as well.

COOLERS

Oil absorbs some of the internal heat developed within the engine. In order for it to do this efficiently, it needs to be cooled. Therefore, an oil cooler is fitted in the oil system of both the wet sump and dry sump systems and is normally cooled by ram air. Some systems incorporate adjustable shutters to control the amount of cooling air entering the cooler and can be automatically positioned in response to signals from an oil thermostat.

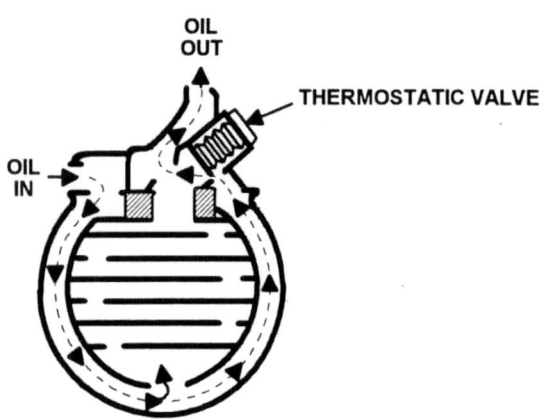
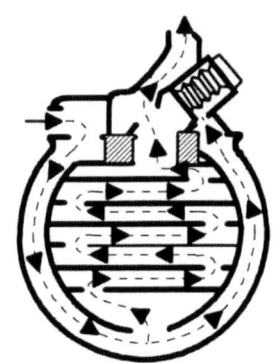

Fig. 4.5

If the oil is overcooled, a condition called **coring** can occur. The oil in the matrix congeals and stops the oil flow. As the oil flow stops, the relief valve opens, returning hot oil back to the engine or tank. A rapid rise in oil temperature is an indication that coring is present. To correct this situation, the cooler needs to be heated to encourage oil flow once more. Closing the shutters or reducing speed and descending to a warmer level can achieve this.

A thermostatic control valve allows cold oil to bypass the core of the cooler. As the oil warms up, a valve diverts the oil through the core so that the air can pick up excess heat. In some cases, the thermostatic valves can also function as a pressure relief valve and are both temperature and pressure sensitive.

FILTERS

Fig. 4.6

Filters are fitted to remove small particles of contamination passing through the lubricating system that can result in damage to the pump and bearings. Most piston engines use the full flow oil system, where all the oil passes through a pressure filter located downstream of the pressure pump. In a dry sump system, scavenge filters are normally located in the scavenge lines from the bearings and pumps. If the filter becomes blocked, engine failure is likely, so the filter has a bypass to allow unfiltered oil to be supplied to the engine. A strainer, usually a coarse mesh type filter, is in the oil sump of a wet sump engine to prevent contamination from entering the system.

PRESSURE GAUGE

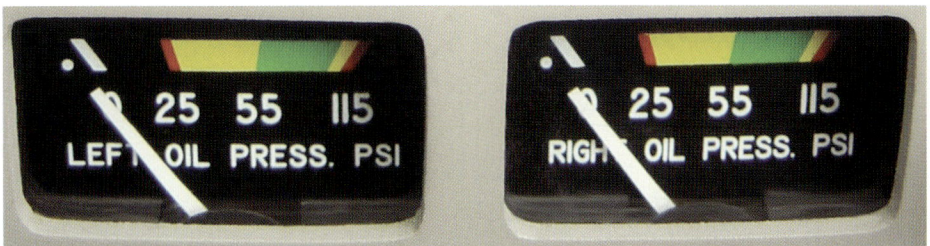

Fig. 4.7

This is the most important gauge for satisfactory engine operation. Should the oil pressure fail, bearing failure occurs very quickly. A green arc on the face of the gauge shows the normal pressure range, a yellow arc for the caution range, and a red line for maximum oil pressure. Oil pressure should register on the gauge within 30 seconds of the engine starting or slightly longer on a cold day. Should the oil pressure not register within this time, shut the engine down.

TEMPERATURE GAUGE

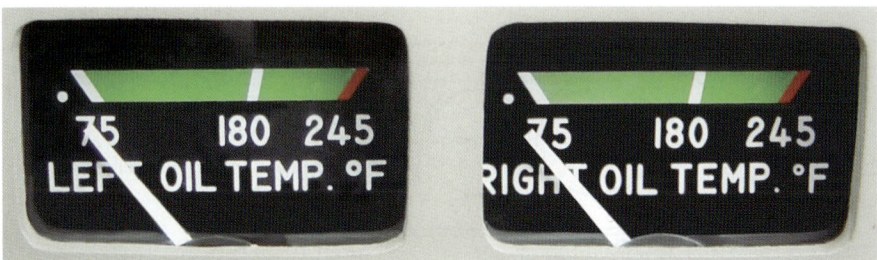

Fig. 4.8

This gauge shows the oil temperature taken after the oil cooler and has a green arc for the normal temperature range and a red line for maximum temperature. The gauges used on light aircraft consist of a sealed Bowden tube filled with a fluid. As the temperature of the lubricating oil rises, it heats the fluid in the tube, causing it to expand. This expansion results in the pointer moving over a temperature scale.

OIL TANK

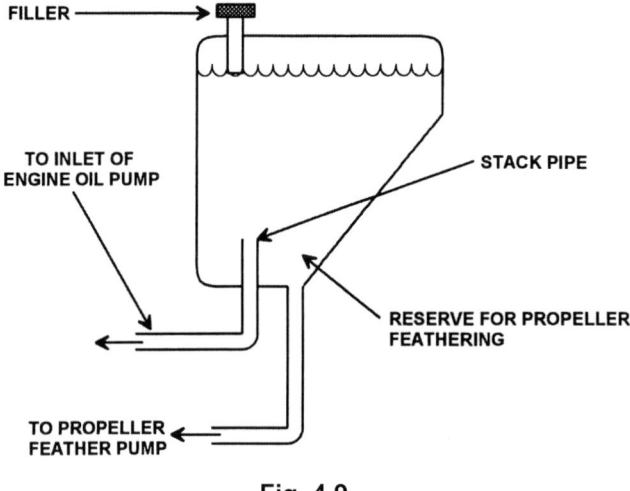

Fig. 4.9

This is only fitted to dry sump systems and is an external tank, which contains the correct amount of oil in order to provide proper circulation and cooling. Engines with constant speed propellers feed the pressure pump through a stack pipe in the tank with the propeller feathering pump outlet at the bottom of the tank. Therefore, should fault occur resulting in a loss of oil, a reserve of oil is retained in the tank for propeller feathering.

COOLING

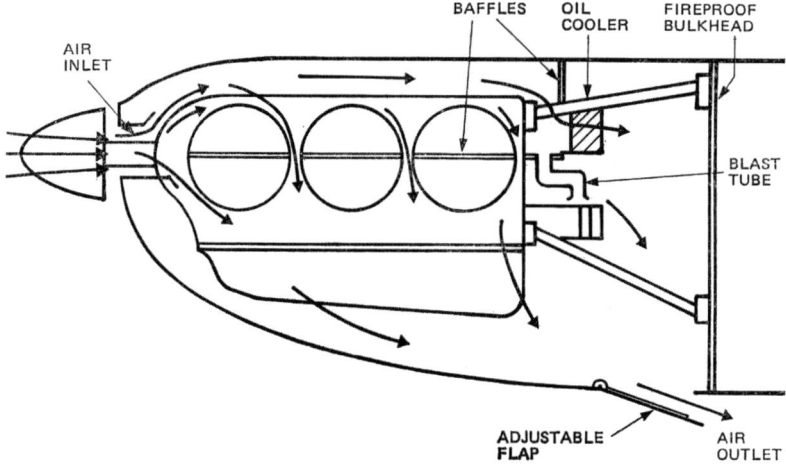

Fig. 4.10

Since only approximately 30% of the heat energy released during combustion is available for power, the remainder of heat has to be removed to prevent engine damage. The majority of the excess heat escapes via the exhaust, whilst the engine absorbs the remainder. The circulating oil absorbs some of this heat, transferring it to air passing through the oil cooler, whilst the engine cooling system removes the remainder.

Air-cooling is the method employed on modern light aircraft and is achieved by passing air over the engine. To improve the cooling effect, the cylinder barrels and cylinder heads are fitted with fins to increase the surface area for efficient heat transfer to the airflow. It is a pressure cooling system where air from the propeller and the aircraft's forward speed is forced through the cowling air inlets to the upper part of the engine, creating a high-pressure region. Baffles then direct the air over the cylinders to the lower part of the engine, which is the low-pressure region. The air exits via a fixed outlet or variable cowl flaps. To direct jets of cooling air over various accessories such as magnetos, alternators, and fuel pumps, blast tubes can be built into the baffles.

COWL FLAPS

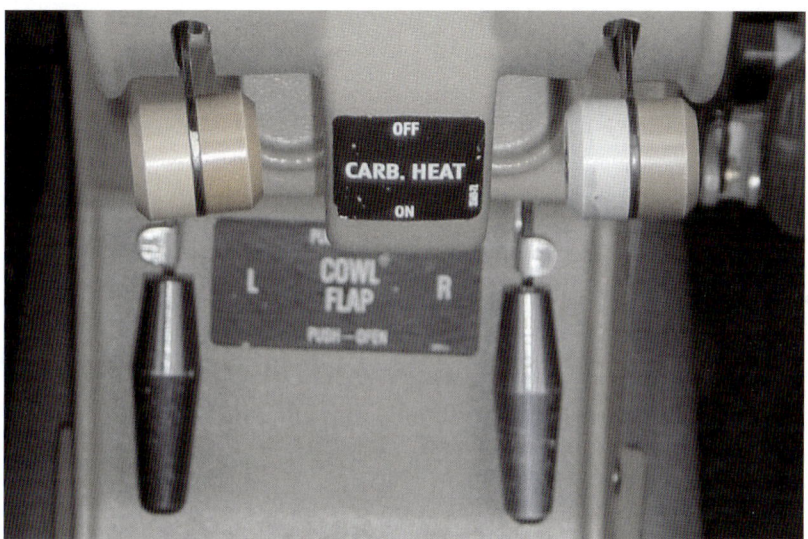

Fig. 4.11

Apart from operating within its limits, the pilot has no direct control over the amount of cooling of most small light aircraft. However, on most large single-engine aircraft, and normally on all twin-engine aircraft, the pilot is able to control the airflow through the engine by the use of cowl flaps as shown above. When open, they allow the maximum amount of air to flow over and around the engine.

When the cowl flap is selected shut, a gap remains that allows a minimum flow. The cowl flaps are normally fully open at take-off and climb, and partially or fully closed during the cruise. The cylinder head temperature gauge records the cylinder temperature, and an adjustment of the cowl flaps to maintain the correct cylinder head temperature is necessary.

CYLINDER HEAD TEMPERATURE GAUGE

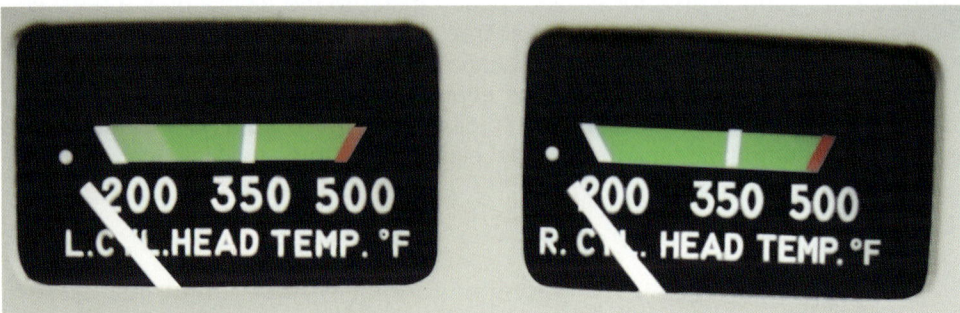

Fig. 4.12

A temperature-measuring probe can either be located in the cylinder head or in the spark plug washer. It operates using the thermocouple principle, previously described for the exhaust gas temperature gauge. The probe is fitted to the hottest cylinder, which is normally one of the rear cylinders.

COMPARISON BETWEEN LIQUID AND AIR-COOLING SYSTEMS

The major advantage of a liquid cooled system is that the engine is kept consistently at a more constant temperature, therefore providing more even cooling. Additionally, the disadvantages are that the system is more complicated and difficult to service. In addition, the engines require water jackets around the cylinders that increase weight. Whilst the air-cooled system is simpler, it contains baffles and deflectors to give very efficient cooling. These systems are fitted to the vast majority of engines.

Chapter 5
Ignition System

INTRODUCTION

If the engine ignition system were part of the aircraft main electrical system, the engine would stop should a complete electrical failure occur. Therefore, an independent aircraft ignition system is required. Not only is it independent of the aircraft electrical system, but it is also duplicated for safety. Modern light aircraft most commonly employ the high-tension magneto system, consisting of at least two magnetos per engine, an integral transformer, and a spark distribution system.

The magnetos perform as self-contained generators, which supply low voltage (low tension) to the integral transformer, which greatly increases the voltage level (high tension). This is fed to the spark distribution system, which delivers the high tension to the correct spark plug at the correct time via the spark plug leads. To satisfy safety requirements, each cylinder contains two spark plugs to ignite the fuel mixture; different magnetos supply each. An ignition switch in the cockpit serves to interrupt magneto operation, as described later.

MAGNETO OPERATION

There are various types of magneto available. The rotating magnet is most common on current piston engines. It may consist of two, four, or eight magnetic poles. Two poles are the most popular.

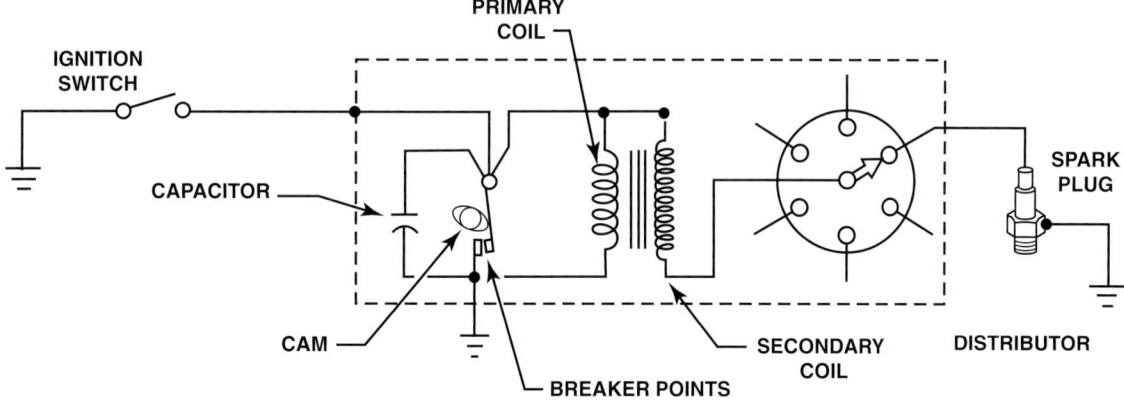

Fig. 5.1

The magneto operates on the principle that a rotating magnetic field created by the rotating magnetic poles induces an electrical voltage in a coil. The voltage produced depends on the strength of the magnetic field and the rate at which the magnetic field changes. The crankshaft drives the magnets as well as the cam and distributor.

PRIMARY CIRCUIT

One side of the magneto circuit is called the primary circuit (see figure 5.1) and contains a coil with relatively few turns of large diameter wire over an iron core. The coil is hard wired to earth on one side and can be earthed by either the contact breaker or the ignition switch on the other side. Spring pressure holds the contact breaker points closed and the action of the cam opens them. The breaker points close when the cam rotates to its low point. A capacitor is wired in parallel with the breaker points and each is in series with the primary coil.

With the ignition switch in the ON (OPEN) position, the primary circuit is only complete when the breaker points are closed. The rotating magnet induces current flow in the primary circuit. This current flow produces its own magnetic field that cuts across the secondary windings, inducing an electrometric force. When the breaker points are opened, the primary circuit is broken and the magnetic field about the primary windings collapses, causing the secondary windings to be cut by the lines of force. This induces a high voltage in the secondary windings. The number of windings in the secondary coil determines its strength.

As the contact breaker opens and closes, arcing occurs, which can cause pitting and corrosion of the points. To prevent this, a capacitor is wired in parallel across the contact breaker points. This capacitor acts as a low resistance path around the breaker points as they open, absorbing the high current until the contact points are too far apart for arcing to occur and ceasing to allow any further current flow through itself as the capacitor becomes fully charged.

Switching the Ignition OFF earths both sides of the Primary so that no voltage can be induced in it. This arrangement has the advantage that if the switch lead breaks, the magneto remains ON. However, should a lead break and touch the aircraft structure, it is earthed and the magneto is OFF.

SECONDARY CIRCUIT

The distributor and spark plugs take their supply from the secondary circuit. This circuit contains the secondary coil (see figure 5.1), which is made up of thousands of turns of thin wire over the top of the primary coil and shares a common earth with it. The distributor takes its timing from the camshaft and distributes the very high output voltage from the secondary coil to the appropriate spark plug.

This consists of a coil, distributor, plug leads, and spark plugs. The coil consists of thousands of turns of very small wire, which wind around the primary coil. The coil is earthed at one end with the primary circuit, and the other terminates at the spark plug, earthing at the spark plug electrodes. The rotor arm of the distributor distributes the high voltage pulses from the secondary coil to the respective plug lead segment inside the distributor. This results in approximately 20 000 volts induced in the secondary coil, terminating as a high intensity spark at the spark plug gap. Turbocharged engine magnetos are usually pressurised from the outlet side of the turbocharger compressor to prevent arcing in the body of the magneto at high altitudes that can occur due to low ambient pressure.

Flight at high altitude creates another problem. The reduced density decreases the electrical resistance of the air, which allows the high voltages induced into the secondary circuit to arc across air gaps within the body of the magneto. To stop this effect, turbocharged engine magnetos are usually pressurised by using air from the outlet side of the turbocharger compressor (deck pressure).

AUXILIARY SYSTEMS FOR STARTING

When the starter is turning the engine during start, the magneto does not produce high voltage required by the spark plugs. In addition, the spark is required to be retarded to prevent kickback. There are various starting devices to accomplish a satisfactory start.

IMPULSE COUPLING

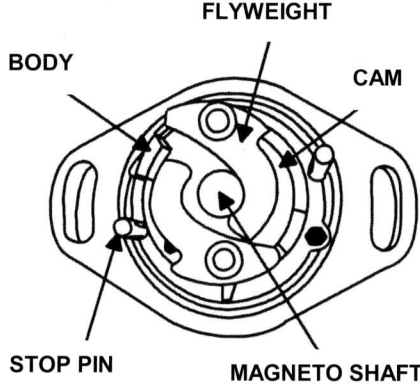

Fig. 5.2

Normally fitted to the left magneto only, this is the most common auxiliary starting system. It is located between the engine and the magneto and consists of a cam plate with two flyweights keyed to the magnet shaft. The impulse coupling body rides over the cam and flyweight assembly and is driven by the engine. A heavy-duty, clock-type spring connects the body and cam plate. Two stop pins are located in the magneto housing. As the starter rotates the engine, the flyweights contact the stop pins, resulting in the magnet and cam plate remaining stationary, which winds up the spring and retards the ignition. After a pre-determined amount of rotation, projections on the coupling body contact the flyweights and release them from the stop pins. The spring now spins the magnet, producing a strong, retarded spark. As soon as the engine starts to fire and accelerate, centrifugal force on the flyweights holds them clear of the stop pins and the magneto reverts to normal operation with the spark in the advanced position.

INDUCTION VIBRATOR SYSTEM (LT COIL)

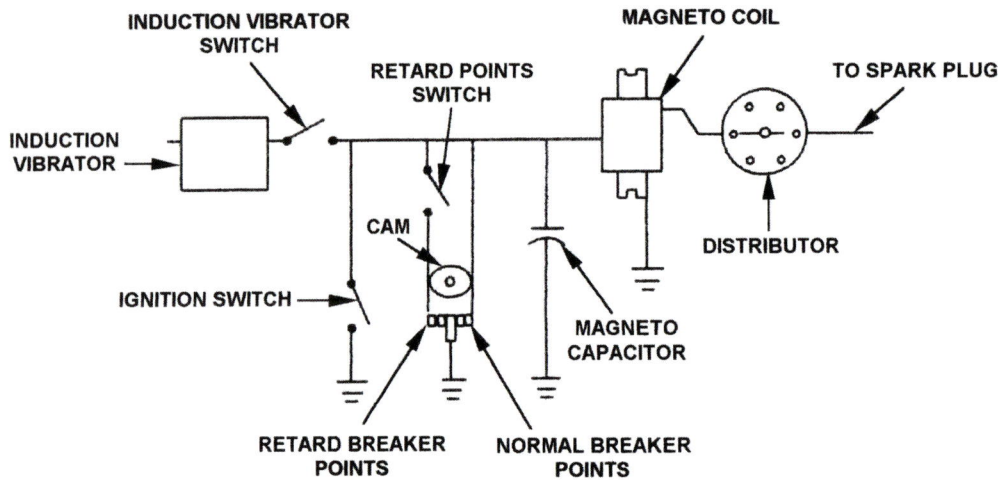

Fig. 5.3

This system is popular on larger engines with six or more cylinders to boost the primary circuit. The vibrator relays direct current pulses to the primary coil, which produces induced high voltage in the secondary coil. This system produces a continuous stream of sparks, sometimes referred to as the **shower of sparks** system, whilst the normal contact breaker points are open. To retard the sparks, a second set of contact breaker points are incorporated which open late, therefore delaying the spark. When the engine is started, pulsating direct current goes to ground through both the normal and the retard points, which are fitted in parallel. The normal points open first at the normal advance position, but as the retard points are still closed, there is no spark at the secondary coil. At the correct position for the starting spark to occur, the retard points open. The only path for the pulsating direct current is via the primary coil to the secondary coil and then to ground at the spark plug, providing a continuous spark until the normal points close. During this time, the distributor rotor is aligned with the electrode for one of the ignition leads to the spark plug. Usually the vibrator is only connected to the left magneto. After start, the system is deactivated, and the magneto functions normally.

HT BOOSTER COIL

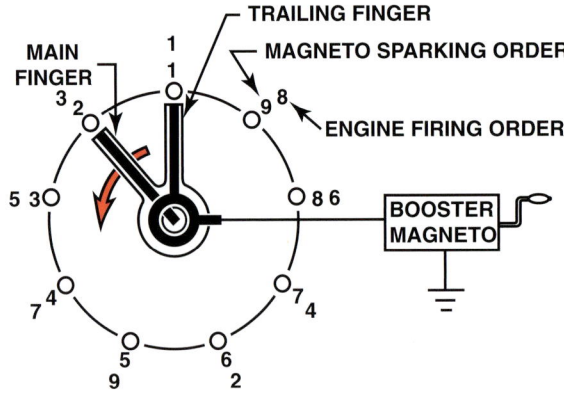

Fig. 5.4

Either battery or ground supply powers this system, which is normally used on older type engines, typically radial engines. When this system operates, it boosts the secondary coil directly, supplying high tension to the correct spark plug via a trailing brush in the distributor. The distributor trailing brush ensures that the spark retards during start. After start, the system is de-activated and the ignition system reverts to normal operation.

PRE TAKE-OFF CHECKS

Carry out before take-off checks on the ignition system to ensure that the magnetos are functioning correctly.

MAGNETO CHECKS

DEAD CUT CHECK

Carry out this check with the engine running at an intermediate rpm, usually 1000 rpm, in accordance with the flight manual. Move the ignition switch from BOTH to L, back to BOTH, then to R, and back to BOTH. When moving the switch to L and R, observe the rpm reduction. A reduction in rpm shows that the magnetos are switching OFF and ON. The purpose of this check is to prevent the possibility of engine damage when carrying out the single ignition check at a higher rpm. If the L magneto is not operating, then switching to L causes the engine to stop. If the switch is turned to BOTH, mixture that has entered the exhaust system may explode and cause damage to the exhaust system. The term **dead cut check** comes from the practice of actually turning OFF both magnetos. However, this procedure is not normally carried out due to the problem identified above in terms of exploding mixture in the exhaust system.

SINGLE IGNITION CHECK

This is carried out at a higher engine speed than the dead cut check to ensure that the ignition system is functioning satisfactorily. Engine speed is increased to the laid down figure in accordance with the flight manual, typically 2000 rpm. Move the switch from BOTH to L and note the drop in rpm, since the left magneto is alive and the right magneto is now dead. Return the switch to BOTH and check that the rpm has returned to 2000 rpm. The drop in rpm should not exceed the figure laid down in the manual, typically 150. Repeat the check, moving the switch from BOTH to R and note the drop in rpm. This time, the right magneto is alive and the left magneto has been turned off. The drop in rpm should not exceed the specified figure. Finally, check that the difference between the two magneto drops when operated individually does not exceed the specified figure, typically 50 rpm. Note that if no rpm drop is observed during the check, this may indicate a switch or wiring fault in the circuit, resulting in permanently alive magnetos.

Chapter 6
Engine Operation

PROPELLER INSPECTION

It is very important to inspect the propeller spinner and back plate for security, damage, and cracks prior to flight. Propeller blades must be free from nicks and other damage that could lead to propeller failure.

BASIC STARTING PROCEDURE

After completing the pre-start checks, carry out engine start in accordance with the procedures detailed in the flight manual. The following procedure is of a general nature, and does not apply to any particular aircraft.

Starting

- ➢ Check fuel selector ON appropriate tank
- ➢ Throttle - ¼ inch to ½ inch open
- ➢ Battery switch - ON
- ➢ Electric Booster Fuel Pump - ON if fitted
- ➢ Mixture - RICH
- ➢ Magneto/Start switch – START position ensuring starter engage light is ON
- ➢ Release the Magneto/Start switch to BOTH position as soon as the engine starts

After the engine starts, check that the starter-engage light is **OUT.** If it remains **ON**, shut down the engine immediately. Oil pressure must register within 30 seconds or slightly longer in cold weather. If the oil pressure fails to register within the prescribed time, stop the engine immediately. Once started, allow the engine to reach its normal operating temperature by running the engine at approximately 1000 to 1200 rpm. Note that prolonged idling or single magneto operation could cause fouling of the spark plugs.

POWER CHANGES

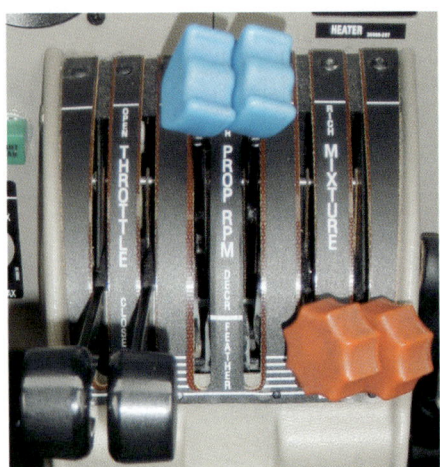

Fig. 6.1

On engines fitted with fixed-pitch propellers, the throttle valve is the only means of controlling power (apart from mixture control). Additionally, constant speed propellers have an rpm lever. Power is a function of rpm and manifold pressure. When making power changes on an engine fitted with a constant speed propeller, adjust both the rpm and the manifold pressure.

When increasing power, the sequence is to adjust the rpm lever first, then the throttle lever. When decreasing power, adjust the throttle lever first and then the rpm lever, ensuring the mixture lever is set to rich before increasing power. This sequence is required in order to avoid detonation and unnecessary stress on the engine.

POWER SETTINGS

The following are various power settings associated with engine operation:

> **Take-Off Power**
> This is the maximum amount of power that the engine can deliver for take-off and is typically limited to a period of one to five minutes.

> **Rated Power or Maximum Continuous Power**
> This is the maximum horsepower obtainable from an engine at specified manifold pressure and rpm as safe for continuous operation.

> **Climb Power**
> Climb power may have a time restriction placed upon it, typically 30 minutes.

> **RPM Limitations**
> Minimum and maximum rpm limits also exist during flight.

> **Critical RPM**
> This is a critical rpm range where continuous operation is not allowed and transit through it should be as quick as possible, since it is an area where there is possibly a resonance or vibration problem. Operating an engine in this range may cause component failure.

GAUGES
RPM INDICATION

An rpm gauge indicates piston engine speed, displaying the revolutions per minute of the crankshaft. It is an indication of engine power and is normally the only indication when operating with a fixed-pitch propeller. The instrument shows the rpm, divided by 100, and the numbers are normally white on a black background. A green arc, with the maximum rpm indicated as a red line, shows the normal operating rpm range. There are two main types of rpm gauges in use:

> ➢ Mechanical tachometer
> ➢ Electrical tachometer

MECHANICAL TACHOMETER

Fig. 6.2

This has a flexible drive from the engine, which is driven at half engine speed and turns a magnet inside a drag cup. The pointer is connected to the drag cup, and the magnetic field induced rotates the drag cup and the pointer.

ELECTRIC TACHOMETER

A three-phase AC generator driven by the engine is the most common type of electric tachometer, where the generator voltage output varies with speed as well as frequency. The instrument consists of a synchronous motor that turns at exactly the same speed as the generator, driving a magnet that operates a drag cup in the same way as in the mechanical type.

MANIFOLD ABSOLUTE PRESSURE GAUGE (MAP)

Fig. 6.3

The MAP gauge indicates the absolute pressure of the fuel/air mixture between the throttle and the inlet valve. It has a sealed capsule and a capsule that is exposed to manifold pressure. The MAP gauge calibration is in inches of mercury where 29.92 in Hg is the standard pressure. In the case of a normally aspirated engine, the highest pressure is at sea level with the throttle fully open. In this condition, the indicated MAP is less than ambient due to losses within the induction system. A turbocharged engine achieves MAP above ambient.

If moisture forms in the MAP gauge line, it could result in erratic operation of the gauge. As a result, a means to drain the moisture is required. This takes the form of a purge valve situated near the instrument inside the cockpit. With the engine running, it is opened forcing air at atmospheric pressure through the instrument drain line into the engine, thereby removing any moisture. For a multi-engine aircraft, there is a purge valve for each engine. The purge valve(s) is operated during engine run-up checks prior to take-off. On engines fitted with a constant speed propeller, the MAP gauge serves as an indication of power.

FUEL FLOW AND PRESSURE GAUGES

Fig. 6.4

Normally, they are direct reading gauges, where the fuel pressure is fed directly to the instrument, and a bourdon tube moves a pointer. Alternative types of flowmeters used in piston engines incorporate a movable vane, which is displaced by the volume of fuel. This displacement is converted into an electrical signal for the instrument in the cockpit and can be calibrated either individually in gallons per hour, pounds per hour, pounds per square inch, or a combination of all of them.

OIL AND TEMPERATURE INDICATIONS

Oil system gauges and temperature gauges are discussed in their relevant sections.

Chapter 7
Piston Engine Performance

INTRODUCTION

The weight of fuel/air mixture consumed by an engine determines the power output. The weight of mixture, controlled by the throttle position, determines the manifold pressure (MAP), and with a constant speed propeller, there is an rpm lever to control engine speed. Therefore, power output is a result of rpm and MAP.

NORMALLY ASPIRATED

This engine breathes normally, taking in air not subjected to supercharging.

THE EFFECTS OF ALTITUDE ON PERFORMANCE

Since air density affects the power output of an engine at a specified MAP and rpm, consider the affects of pressure, temperature, and humidity on the air density. Consider the following terms:

> **Pressure Altitude**
> This is the pressure at standard temperature, requiring no correction unless humidity is a factor.

> **Density Altitude**
> This is pressure altitude corrected for non-standard temperature.

> **Critical Altitude**
> This is the maximum altitude at which an engine maintains a given horsepower output. In other words, an engine may be rated at a particular altitude, which is the highest level where rated power output can be obtained. Superchargers and turbochargers increase the critical altitude.

Since power depends on air density, convert pressure altitude to density altitude for performance calculations. If the temperature at a particular pressure altitude is the same as standard, then no correction for density is required. Figure 7.1 shows a typical correction chart.

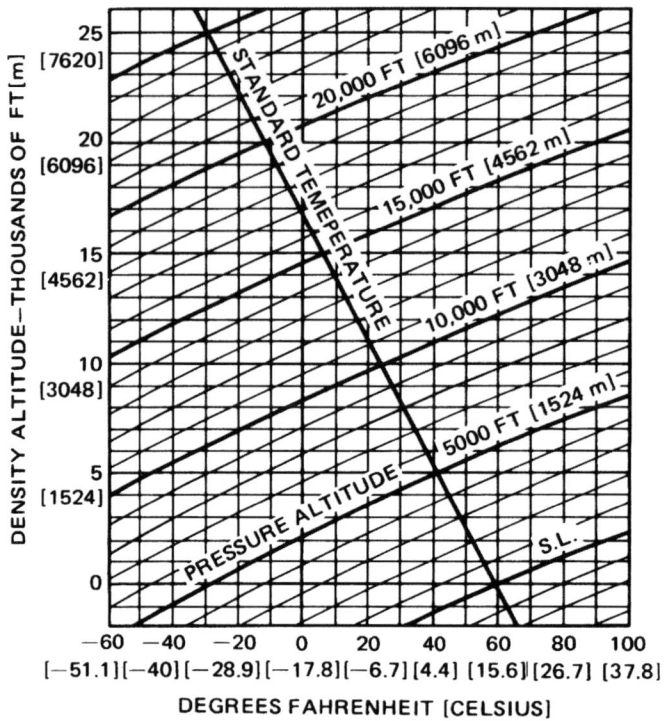

Fig. 7.1

This formula is another method of calculating density altitude:

Density Altitude = Pressure Altitude + (Temperature Deviation x 118.8)

Pressure, temperature, and humidity affect the density of air and hence power output in the following ways. A reduction in pressure reduces density, therefore reducing power output. A temperature reduction increases density, therefore increasing power output. An increase in humidity reduces density and decreases power output. The main factors considered in engine performance are effects of pressure and temperature.

Pressure and temperature decrease with altitude. The pressure reduction reduces density and the temperature drop increases density. This interrelationship results in the temperature reduction offsetting the pressure drop somewhat; however, the drop in pressure has more effect than the temperature drop, resulting in decreased air density and decreased power with altitude.

When climbing at constant MAP and rpm, since the pressure is constant, the reduction in temperature increases the density of the air, resulting in a gradual increase in power. Constant MAP is maintained by opening the throttle during the climb until the throttle is fully open, after which any further increase in altitude results in a reduction in power.

When the throttle is fully open it corresponds to a particular height, called the **full throttle height (FTH)**, for that power setting. Each power setting in terms of MAP and rpm has its own FTH. The lower the power, the higher the FTH and vice versa, as illustrated in figure 7.2.

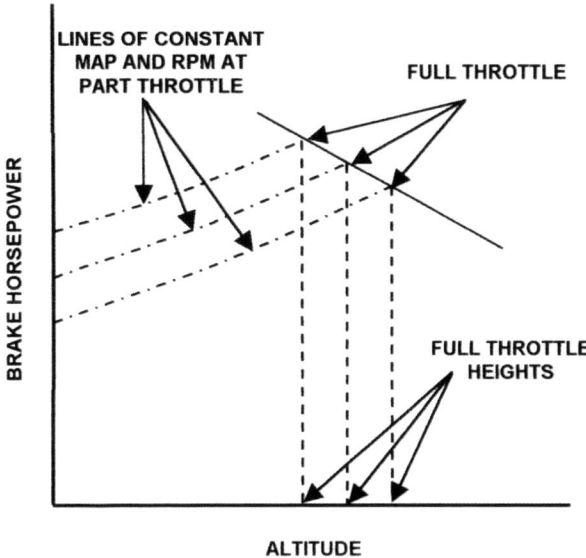

Fig. 7.2

In addition, while gaining altitude, the reduction in pressure causes a reduction in **exhaust backpressure**, which is the resistance of the exhaust gases leaving the exhaust. Therefore, the reduction in exhaust backpressure results in the exhaust gases leaving the exhaust more freely. This results in improved scavenging of the exhaust gases and induction of the mixture, improving volumetric efficiency. The combined effect of reducing temperature and increasing volumetric efficiency increases power whilst climbing at constant MAP and rpm. However, while climbing at full throttle and constant rpm, a reduction of MAP and power results due to the decreasing density. The effect of this is illustrated in figure 7.3. Note that a constant speed propeller that has its blade angle varied by a propeller governor maintains constant rpm. This operation is described in the propeller chapter.

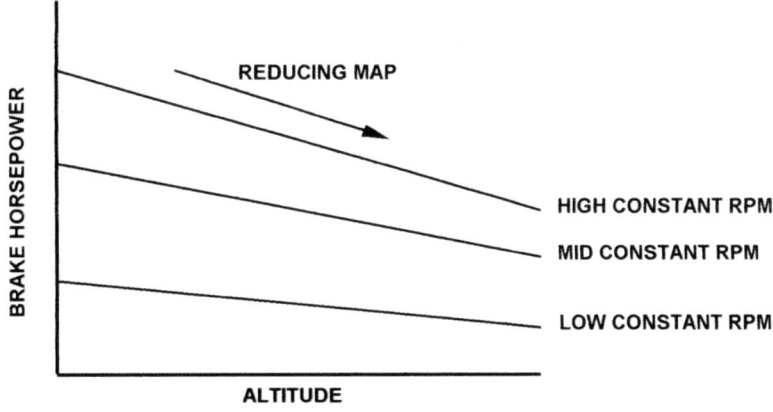

Fig. 7.3

Performance charts used to determine power output use standard ISA conditions, therefore make corrections to power output by subtracting 1% for every 6°C above standard and vice versa.

ADDITIONAL FACTORS THAT AFFECT ENGINE PERFORMANCE

There are additional factors to take into account that affect engine performance, but that are not normally considered in engine performance charts. The description of some of these follows.

RAM AIR PRESSURE

This is the pressure at the carburettor intake due to forward speed of the aircraft. It has the effect of supercharging the air, and has the effect of increasing engine power over that achieved under standard conditions of pressure, temperature, and rpm.

HUMIDITY

High humidity lowers air density, but is not usually considered in engine performance charts. It is advisable to consider the affect humidity has on engine power if other factors are critical when taking off from a short field.

CARBURETTOR AIR TEMPERATURE

If air temperature increases, a reduction in density results, and subsequently the weight of air taken into the engine decreases. If the temperature is too high, detonation occurs. With a turbo-charger or super-charger, rather than the intake temperature, observe the compressor outlet temperature.

CRUISE CONTROL

This is the adjustment of engine controls to obtain the desired results in economy, range, or flight time. Because an engine uses more fuel at high power settings than at low settings, maximum speed and maximum range or economy is not achieved with the same power settings. Should a maximum distance flight be required, conserve fuel by operating at a low power setting. Conversely, if maximum speed is required, use a maximum power setting, which decreases the range capability.

Chapter 8
Piston Engine Fuel Injection

FUEL INJECTION SYSTEM

There are different types of fuel injection systems available, however the following concentrates on the direct fuel injection system fitted to the more powerful engine types and usually used when a turbocharger is installed. One of the disadvantages of a carburettor is that an imbalance in mixture ratio between the cylinders can occur. This is due to problems associated with the flow characteristics through the inlet manifold caused in part by differing manifold lengths. In the direct fuel injection system, fuel is evenly distributed to the inlet manifold and sprayed continuously into the inlet ports, thereby overcoming this problem by delivering an even mixture ratio to the cylinders. In fuel injection systems, the only thing the throttle valve controls directly is the airflow into the inlet manifold.

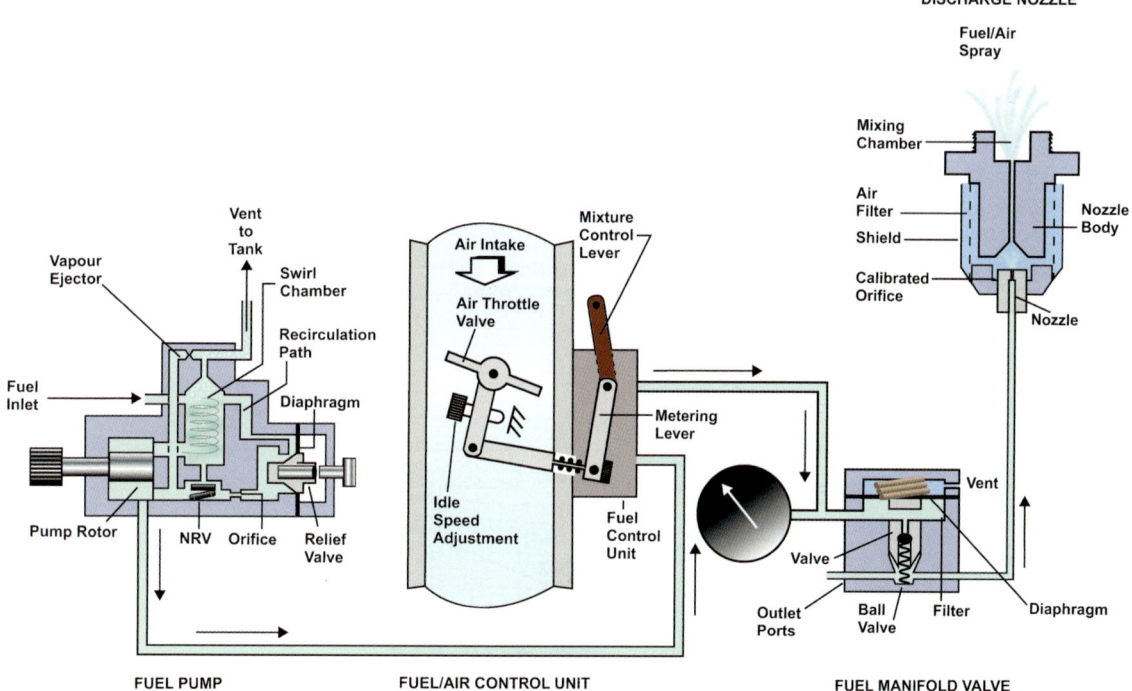

Fig. 8.1

The fuel injection system described (figure 8.1) is a multi-nozzle, continuous flow system that controls fuel flow to match engine requirements. Any change in the air throttle position, engine speed, deck pressure (deck pressure refers to the pressure between the compressor outlet pressure of a turbocharger and the air throttle valve), or any combination of these causes changes in fuel pressure in the correct relationship to engine requirements. A manual mixture control and a fuel flow gauge indicating metered fuel flow are provided for precise leaning at any combination of altitude and power setting. As fuel flow is directly proportional to metered fuel pressure, settings can be predetermined and fuel flow accurately predicted and controlled.

INJECTION PUMP

This is a positive displacement rotary vane-type pump driven by the engine. The pump provides a greater capacity than required by the engine under all running conditions, thus requiring a recirculation path. A spring loaded diaphragm-style relief valve, which also acts as a pressure regulator, is provided in the body of the pump. Pump outlet fuel pressure passes through a calibrated orifice before entering the relief valve chamber, thus making the pump delivery pressure proportional to engine speed. Fuel enters the pump assembly at the swirl chamber of the vapour separator. Any vapour present separates out and rises to the top of the swirl chamber where it is drawn off by means of the vapour ejector. The ejector is a small venturi positioned in the fuel return line, hence a mixture of fuel and vapour returns to the aircraft fuel tank.

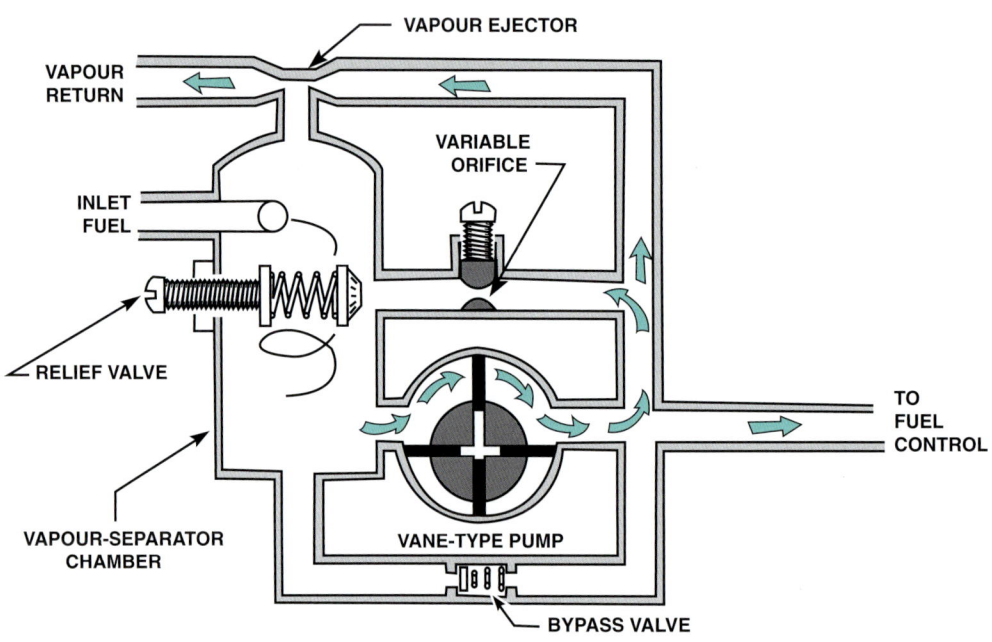

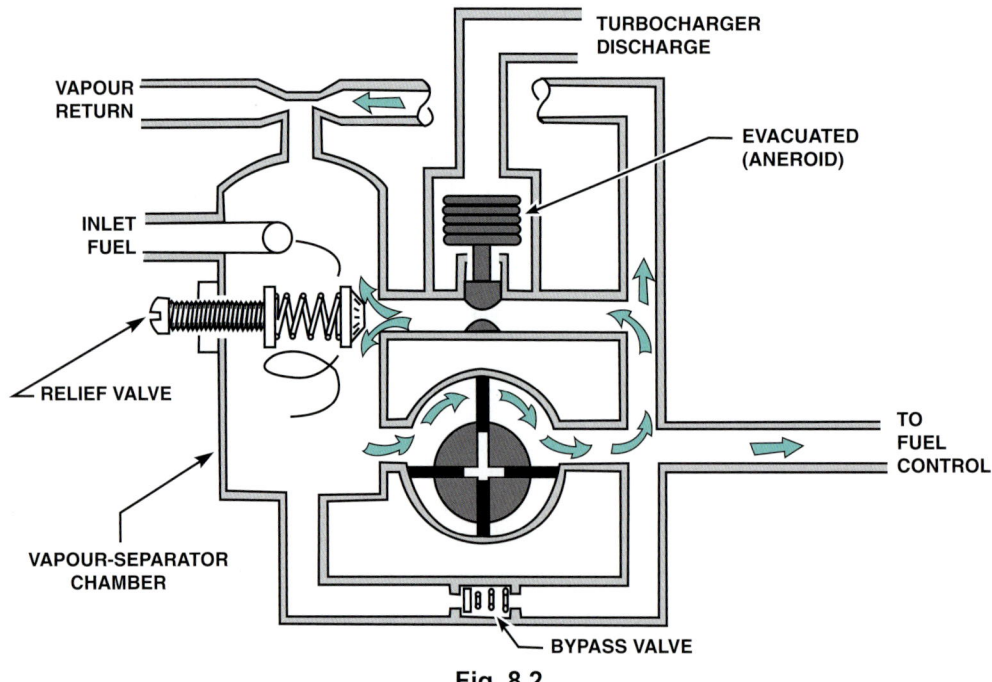

Fig. 8.2

ELECTRIC PUMP

An electric fuel pump is fitted to prime the engine during start, suppress vapour during aircraft operation in high ambient temperature or altitude, and for continued engine operation in the event of engine-driven pump failure.

FUEL/AIR CONTROL UNIT

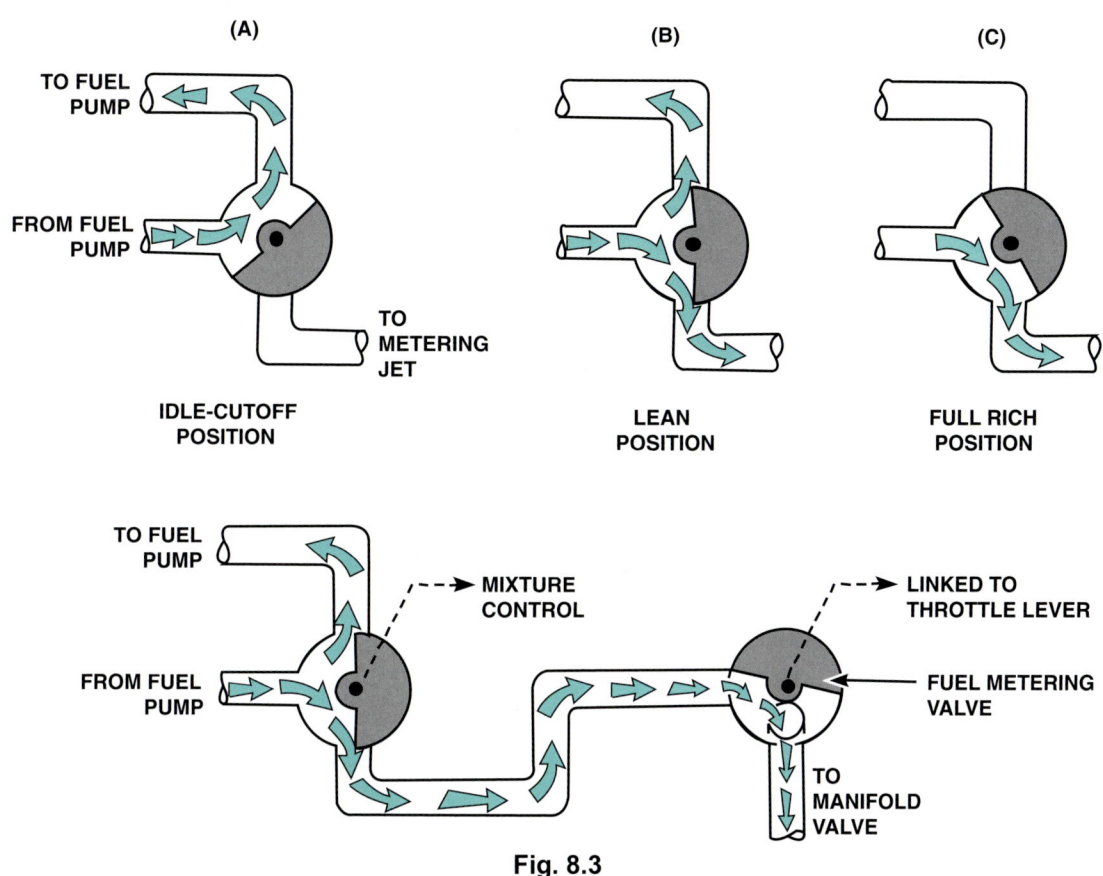

Fig. 8.3

The function of the fuel/air control unit is to control the amount of air admitted into the inlet manifold and to meter the fuel in the correct quantity to provide the correct fuel/air ratio. The location of the air throttle valve is in the intake manifold and is mechanically linked to the fuel-metering valve housed in the fuel control unit. Movement of the air throttle valve thus determines the position of the fuel metering valve and ensures that the correct amount of fuel is delivered to the fuel manifold valve. Also housed in the fuel control unit is a mixture control. Fuel from the pump is delivered to the fuel-metering valve via the mixture control valve. When 'cut-off' is selected, all fuel now passes back to the inlet side of the pump. The pump always delivers more fuel than the engine requires during normal operation, with the least amount of fuel returning to the inlet side of the pump at full throttle and full rich positions. The mixture control lever operates the mixture control valve and the throttle lever operates the air throttle valve and metering valve.

FUEL MANIFOLD VALVE

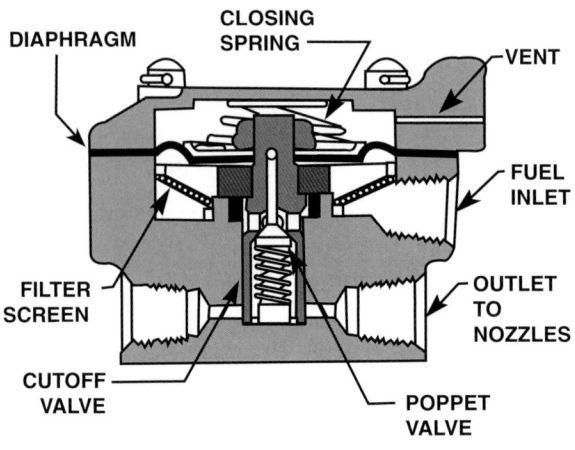

Fig. 8.4

The manifold valve receives fuel from the metering unit. When the fuel pressure reaches a pre-determined pressure, the valve opens allowing the fuel to flow under pressure to each injector nozzle. The manifold valve also serves to provide a positive cut-off of fuel to the cylinder when the engine shuts down. The manifold valve is normally located centrally on top of the crankcase and is referred to as the **spider**. An air vent line is taken from the manifold valve to exit below the engine nacelle; if fuel issues from this vent, the manifold valve is faulty.

INJECTOR NOZZLES

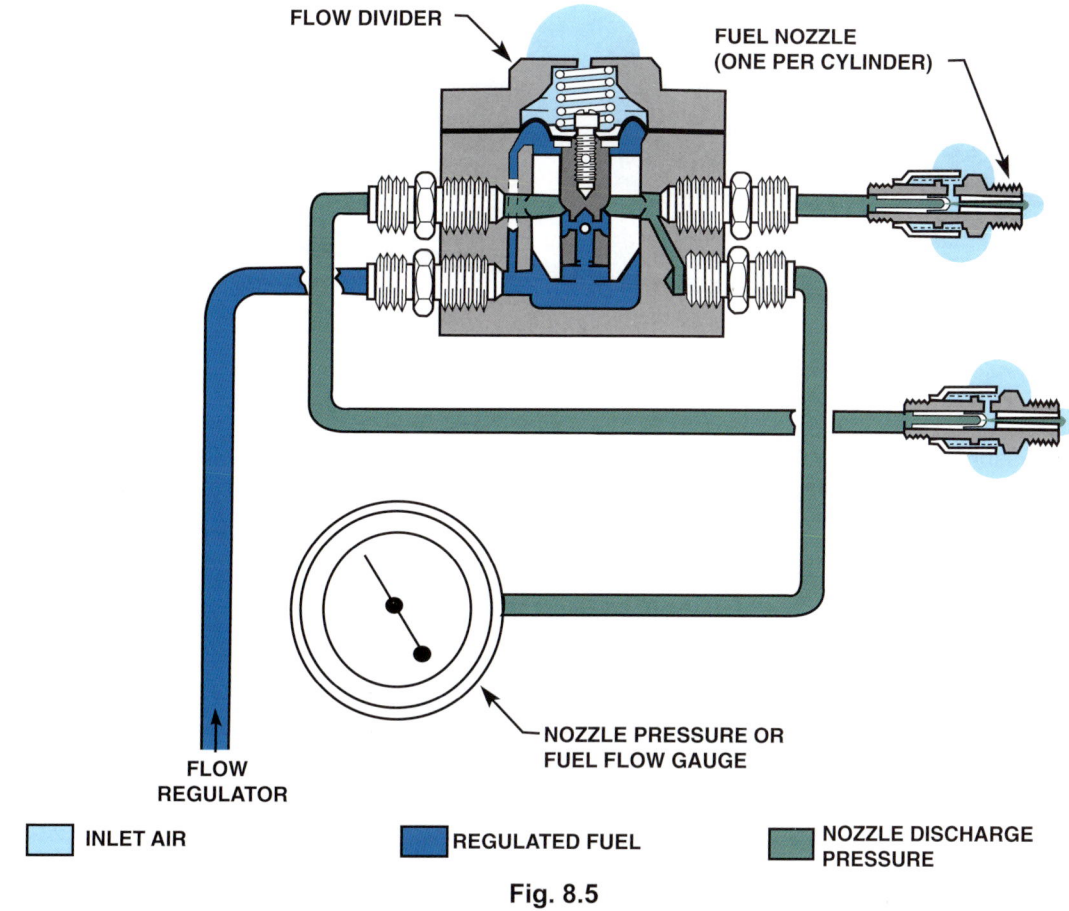

Fig. 8.5

There is one nozzle per cylinder. These nozzles are of the **air bleed** type that sprays fuel directly into the inlet port of the cylinder. When the engine is running, the flow through the nozzle is continuous and enters the cylinder combustion chamber when the inlet valve opens. Since the size of the fuel nozzles is fixed, the pressure applied determines the amount of fuel flowing through them. Therefore, fuel flow may be accurately determined by measuring the pressure at the manifold valve. In some installations, deck pressure pressurises the nozzles to ensure a positive fuel flow at altitude.

FUEL PRESSURE GAUGE

Fig. 8.6

The measure of fuel pressure is in the metered fuel pressure line, which supplies the gauge in the cockpit. The gauge reading is proportional to fuel flow because the mixture ratio is dependent on the pressure of the fuel passing the metering valve.

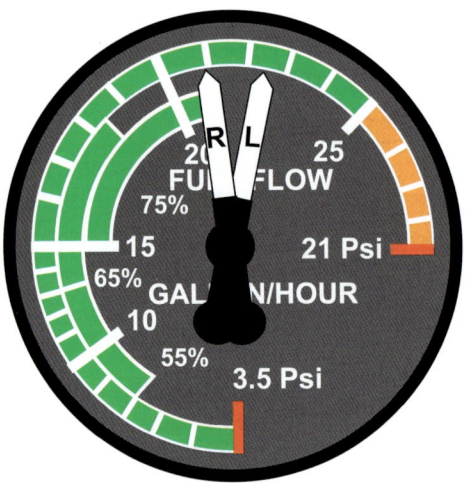

Fig. 8.7

SYSTEM ADVANTAGES

Apart from the advantages previously mentioned as regards the mixture ratio, injection systems provide improved control over the mixture ratio. They are free of icing caused by the vaporisation of the fuel, therefore making it unnecessary to use carburettor heat except under the most severe atmospheric conditions; however, an alternate air control is provided which is described below. They also reduce maintenance problems.

ALTERNATE AIR CONTROL

An alternate air control is normally mounted either on the instrument panel or on the lever quadrant. Its purpose is to admit air into the induction manifold should a blockage occur. The alternate air control can incorporate an automatic alternate air door in the induction system, which opens when induction system suction reaches a pre-determined value. The suction may be caused by impact ice on the face of the air filter or by a collapsed filter.

Chapter 9
Piston Engine
Power Augmentation Systems

INTRODUCTION

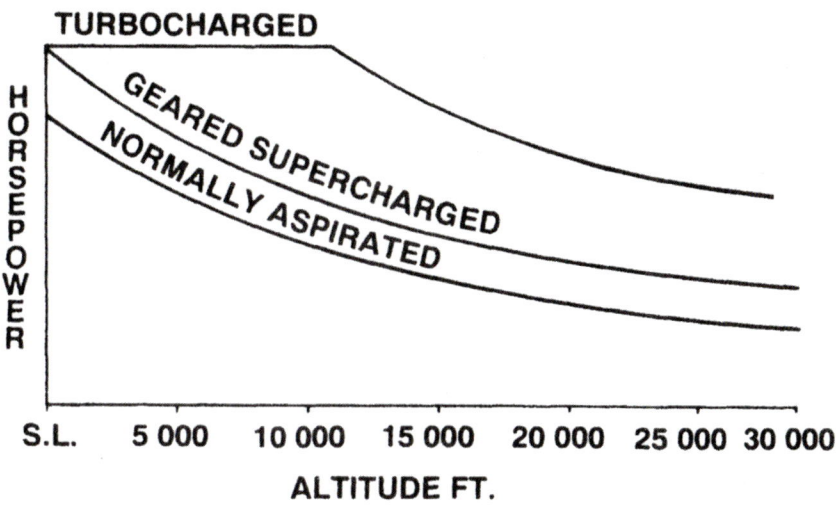

Fig. 9.1

The power developed by a piston engine depends upon the weight of mixture burnt in the cylinders in a given time. Since a normally aspirated engine produces maximum power at sea level, it follows that a loss of power results as altitude increases due to the decrease in pressure and density. To overcome this loss of power, increase engine power for take-off, or to maintain sea level conditions at altitude, the intake manifold pressure and, therefore, the weight of the mixture entering the cylinders must be maintained or increased. Methods referred to as **supercharging** or **turbocharging** can achieve this. Both systems use a compressor consisting of a rotating impeller and a static diffuser.

> **Superchargers**
> These are internal devices, driven by the crankshaft via a gearbox, that compress air/fuel mixture delivered from the carburettor.

> **Turbochargers**
> These are external devices that use a turbine to drive the compressor and usually compress air only. The fuel is typically injected directly into the cylinder.

COMPRESSOR

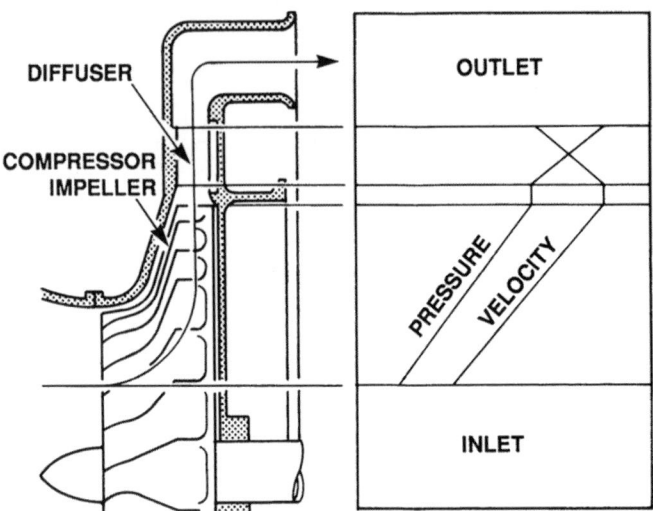

Fig. 9.2

The compressor of a supercharger or turbocharger takes the form of a centrifugal impeller, is comparatively light and reliable, and can run at high speed, enabling it to handle large quantities of air. The centrifugal impeller is, effectively, a fan. When rotated at high speed, it causes the air between the vanes to fly outward due to centrifugal force. The air receives kinetic energy as it flows outward between the vanes, and since the vanes are divergent in cross section, some of this energy converts into pressure energy. The amount of pressure the impeller gains depends on the diameter of the impeller, rotational speed, and the shape of the vanes. Air leaves the impeller with a large tangential and radial velocity and passes into the diffuser. The diffuser consists of divergent fixed vanes, which reduce velocity whilst increasing the pressure. Figure 9.2 illustrates the velocity and pressure changes through the impeller and diffuser vanes. Temperature rises as the air pressure increases.

SUPERCHARGER

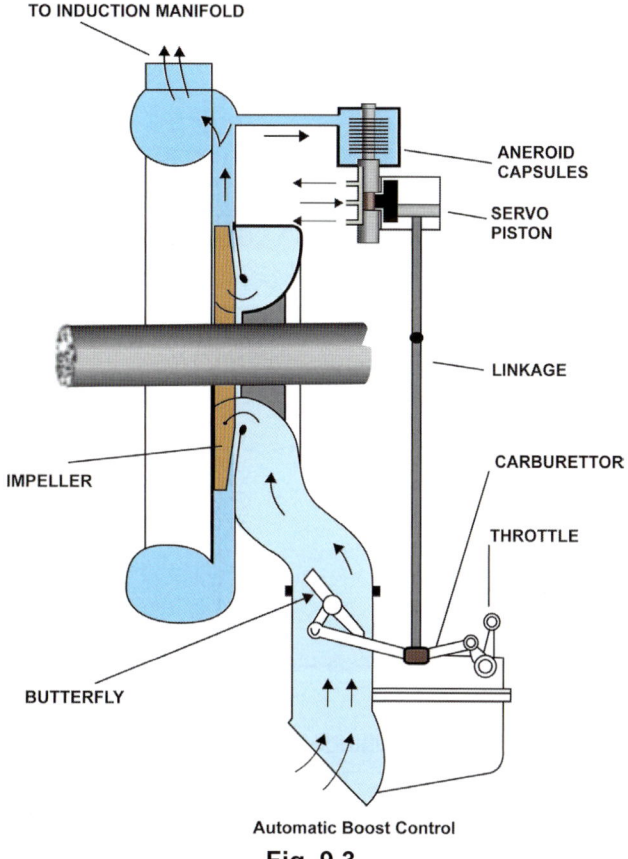

Automatic Boost Control

Fig. 9.3

In a typical supercharger, the engine crankshaft drives the impeller via a gear train. Air entering the intake is drawn through a carburettor where it mixes with fuel. The throttle valve determines the amount of fuel/air mixture delivered to the eye of the impeller. The mixture receives an increase in pressure as it passes through the impeller and a further increase in the diffuser before entering the induction manifold. To prevent excessive manifold pressures, an aneroid unit, sensing manifold pressure, is coupled to the throttle linkage though an oil-operated servo piston in such a way that when the aneroid capsule compresses, the throttle butterfly valve is partly closed, controlling MAP to a safe value.

SUPERCHARGER AND TURBOCHARGER PERFORMANCE

In the case of a normally aspirated engine, air density decreases during the climb. This results in a reduction of the weight of mixture pulled into the cylinders, therefore reducing power. By comparing identical engines, one fitted with a supercharger and one without at the same manifold pressure and speed, the normally aspirated engine produces more power at sea level. Refer to figure 9.4. This is due to the power required to drive the supercharger.

The power output of a supercharged engine, when climbing at a constant manifold pressure, increases due to decreased atmospheric air temperature. This, in turn, increases air density and the weight of mixture entering the cylinders.

A reduction of backpressure occurs due to the reduced atmospheric pressure acting on the exhaust gas leaving the engine. This results in improved cylinder scavenging. The combination of these two factors increases the volumetric efficiency of the engine.

Figure 9.4 also shows that power can be increased for take-off. However, this take-off power setting usually has an imposed time limit of typically 5 minutes.

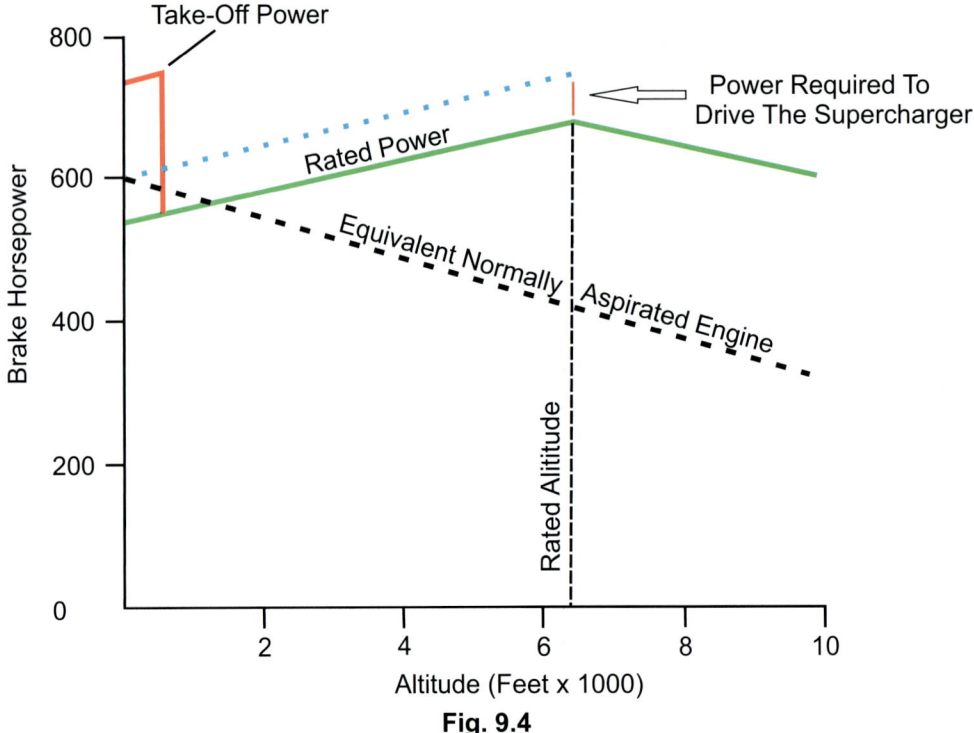

Fig. 9.4

Constant manifold pressure is only maintained by progressively opening the throttle during the climb. Once the throttle is fully open, the **full throttle height** position is reached that corresponds to a particular altitude. After this point, any further increase in altitude results in a reduction in power. If operating at **rated power**, where the definition of rated power is the maximum horsepower at which an engine is approved for continuous operation, the full throttle height position is referred to as **rated altitude** for a supercharger and **critical altitude** for a turbocharger.

In the case of a turbocharger with a fixed waste gate system, using the throttle to maintain the required manifold pressure up to full throttle height achieves full throttle height. With an automatic waste gate system, the throttle, depending on the manifold pressure required, can be positioned up to fully open. The waste gate automatically moves toward closed in the climb to maintain the selected manifold pressure, thus achieving the same effect. Once fully closed, the full throttle height is reached for that power setting, and any further increase in altitude results in a power reduction. The following paragraphs describe operation of the various turbocharger waste-gate systems.

Note that each power setting would have its own full throttle height as illustrated in figure 9.5. In the case of normally aspirated and turbocharged engines, any reduction in power setting, either manifold pressure or rpm, results in an increase in full throttle height. For a supercharger, reducing rpm whilst maintaining manifold (Boost) pressure results in a reduction in full throttle height.

This is due to the mechanical arrangement of the supercharger, since reducing rpm decreases the speed of the compressor. Therefore, in order to maintain the required manifold pressure, the compressor must compress more mixture at the slower speed and the throttle must be opened. In doing so, the mechanical forward movement of the throttle decreases, which results in a lower full throttle height.

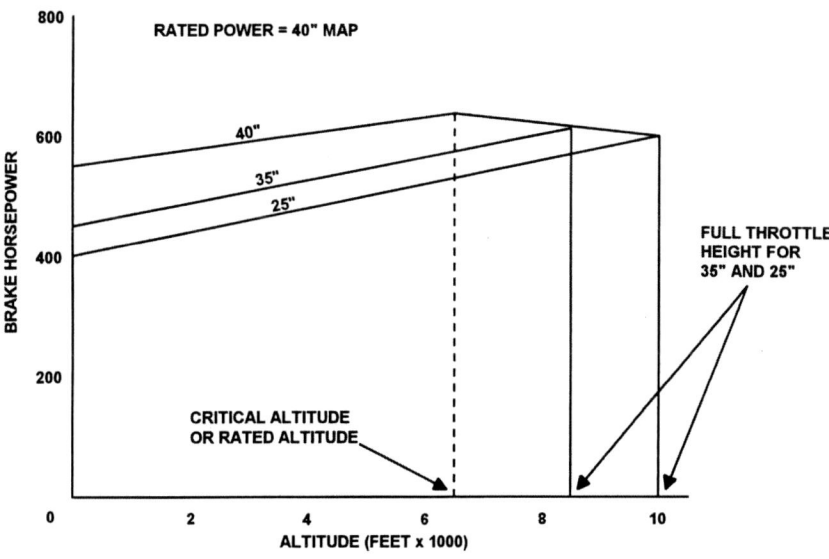

Fig. 9.5

TURBOCHARGER

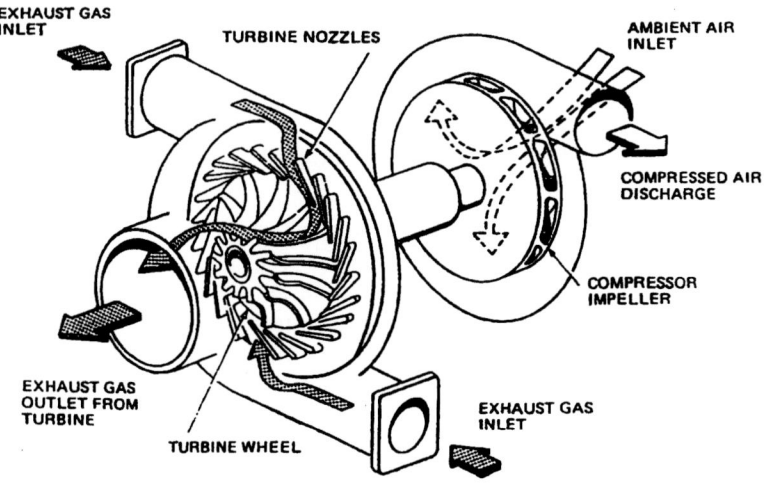

Fig. 9.6

A turbine located in and driven by the exhaust gas, sometimes referred to as either an **external supercharger** or **turbo-supercharger**, drives a turbocharger's compressor. In this arrangement, the throttle valve sits after the compressor and in the induction manifold.

Compressor

Turbine

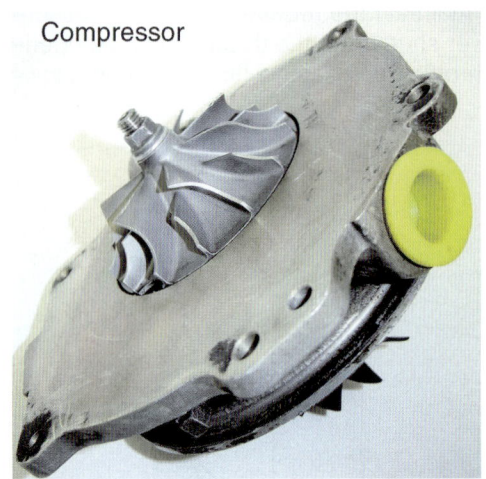

Fig. 9.7

The primary components of a turbocharger are:

> ➢ A turbine and compressor connected by a common shaft.
> ➢ A waste gate that is fitted in parallel to the turbine and controls the amount of exhaust gas that bypasses the turbine.

There are two types of turbocharger to increase power either at altitude or sea level:

> ➢ **Altitude turbochargers** are turbochargers designed to maintain sea-level pressure with increasing height.

> ➢ **Ground-boosted turbochargers** are designed to increase MAP above sea level pressure at sea level standard conditions.

Terminologies associated with the pressures in turbocharger operation include:

> ➢ **Discharge pressure** or **deck pressure** is the air pressure between the compressor outlet and the throttle valve.

> ➢ **Manifold pressure (MAP)** is the air pressure between the throttle valve and the inlet valve. This pressure is always less than compressor discharge pressure due to the pressure drop across the throttle valve, and is the pressure indicated in the cockpit.

While the throttle valve controls the manifold pressure entering the engine, the waste gate controls the amount of exhaust gas that bypasses the turbine. As the turbine and compressor mount on the same shaft, controlling the bypass controls the compressor speed and discharge outlet pressure to maintain the desired manifold pressure.

Depending on the throttle setting, the MAP in a supercharged/turbocharged engine may be greater than or less than atmospheric air pressure. The primary purpose of increasing MAP above atmospheric pressure is to increase take-off power and to maintain rated power at altitude. Increased MAP increases power output by increasing the weight of mixture delivered to the cylinders and by increasing combustion chamber pressure. Note that the fuel octane rating and the stress that the engine can tolerate limit MAP.

WASTE GATE

The waste gate is located parallel to the turbine and controls the amount of exhaust gas bypassing the turbine and consequently the turbine's speed, the speed of the compressor, and thus the compressor discharge pressure. Waste gates may be fixed, operated manually, or operated automatically.

FIXED WASTE GATE

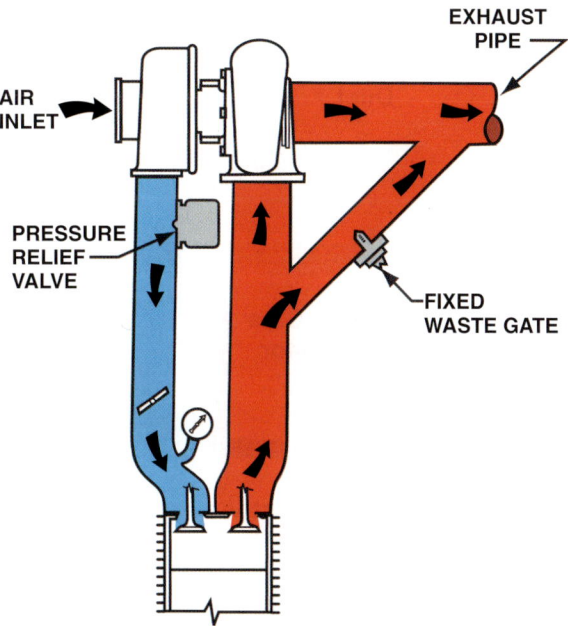

Fig. 9.8

Figure 9.8 illustrates a fixed waste gate. This is a ground-adjusted device. Only a small portion of exhaust gas bypasses the turbine. Although it is a very simple system, the turbocharger is always working and its maximum output is limited. When operating this type of turbocharger, the throttle controls the MAP. Take care not to over-pressurise (overboost) the engine.

OVERBOOST PROTECTION

To guard against overboosting the engine, a manifold pressure relief valve is normally fitted to most systems and is located in the induction manifold as near to the compressor outlet as possible. The valve is spring-loaded and senses deck pressure. When an overboost condition occurs, the spring compresses and the excess pressure vents overboard until the deck pressure falls to within normal limits. A red warning light usually illuminates to indicate an overboost condition.

MANUALLY-OPERATED WASTE GATE

There are various ways to adjust a manual waste gate. In its simplest form, it is sometimes referred to as a variable waste gate, and can move from fully open to fully closed in flight by operation of a cockpit lever. Moving the lever rearward opens the waste gate, therefore diverting the maximum amount of exhaust gas away from the turbine. At take-off, the throttle is set to fully open and the waste gate is set fully open. During the climb, MAP drops. Therefore, the waste gate lever is moved forward to close the waste gate in order to maintain the required MAP until the waste gate reaches fully closed. Any further increase in altitude now results in MAP decreasing, therefore decreasing power.

WASTE GATE ACTUATOR

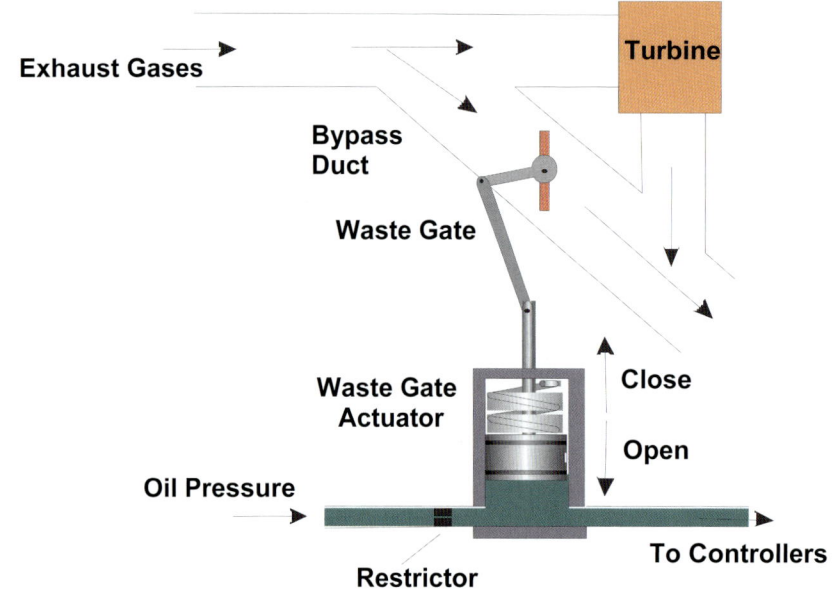

Fig. 9.9

A waste gate actuator, via a mechanical linkage, operates an automatically-operated waste gate system. The waste gate actuator is a spring-loaded piston type actuator. When the engine is stationary, the spring holds the waste gate open. On engine start, oil pressure from the engine oil system flows via a restrictor into the actuator and acts on the piston compressing the spring. Since the piston connects to the waste gate by a linkage, it starts to close and directs more exhaust gas over the turbine. From the waste gate actuator, oil flows to a controller and then returns to the engine. The controller regulates the flow of oil through it, which in turn controls the oil pressure acting on the piston, and hence the waste gate position.

Increasing oil flow through the controller creates a pressure drop on the actuator side of the restrictor, allowing spring force to open the waste gate. Reducing the oil flow through the controller increases the oil pressure, closing the waste gate. Various controllers may be fitted to control the position of the waste gate to cater for all engine requirements and prevent overboosting. A description of these follows.

ABSOLUTE PRESSURE CONTROLLER

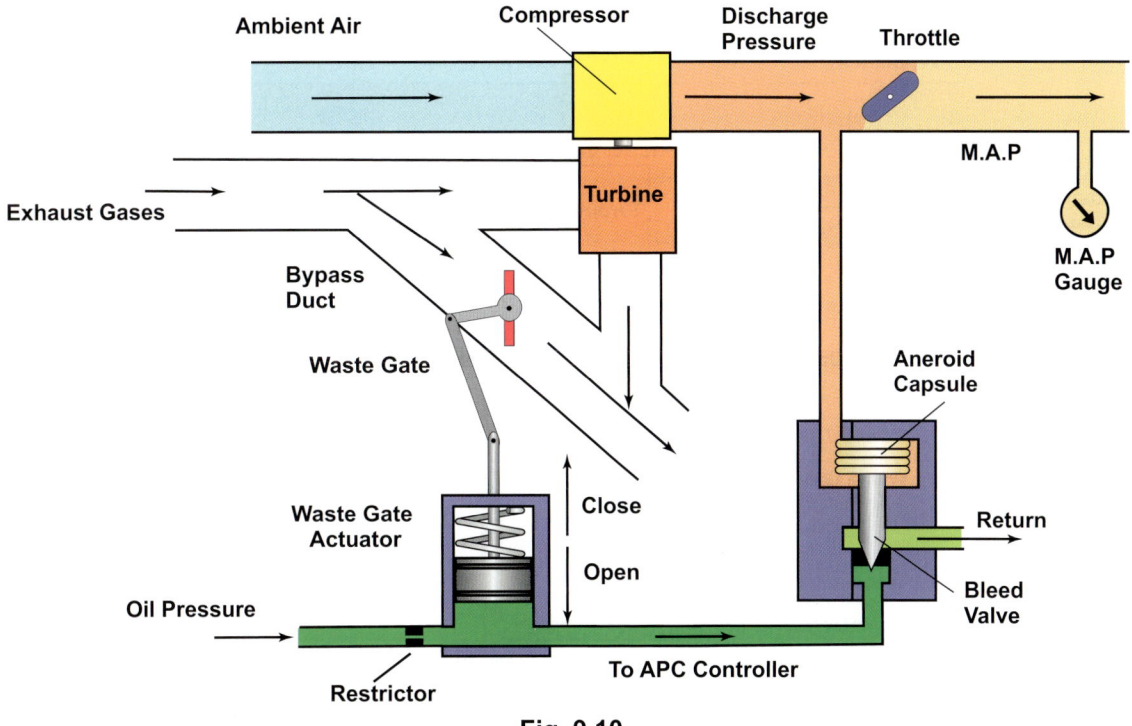

Fig. 9.10

By design, the absolute pressure controller keeps the engine operating within its designed limits, thus preventing the turbocharger outlet pressure from exceeding a specified maximum value. Full oil pressure applies to the waste gate actuator at low power settings; closing the waste gate and diverting all the gas through the turbine. Opening the throttle increases engine speed, and more gas passes through the turbine, resulting in increased turbine speed and higher MAP. Upon reaching the controlling outlet pressure, the capsule contracts, which bleeds oil and opens the waste gate. The waste gate is almost fully open at high power settings and low altitude. As the aircraft climbs, the waste gate closes until becoming fully closed at the critical altitude. Any further increase in altitude reduces MAP and power.

VARIABLE PRESSURE CONTROLLER

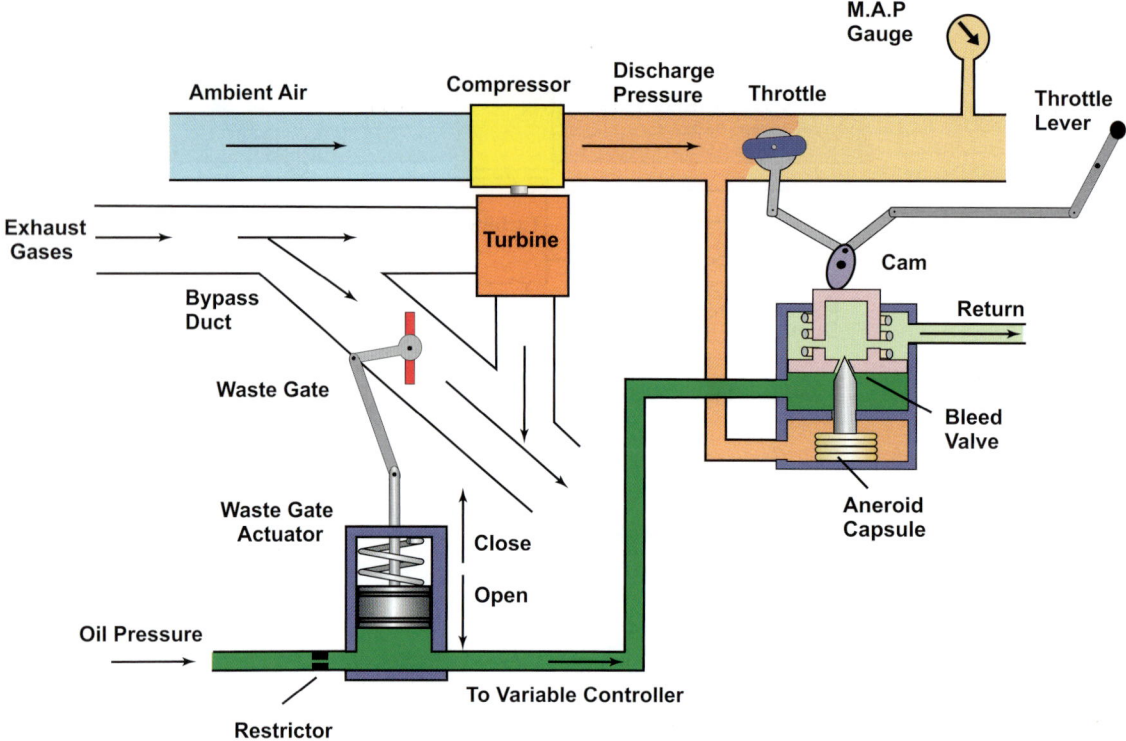

Fig. 9.11

This controller is similar to the absolute pressure controller, except it has a variable datum. A cam operated by a linkage to the throttle control lever adjusts the datum of the bleed valve; therefore controlling the amount the waste gate opens. This results in a manifold pressure related to the power selected by the throttle lever. The capsule still controls the maximum discharge pressure in the same manner as the absolute pressure controller.

DUAL UNIT CONTROLLERS

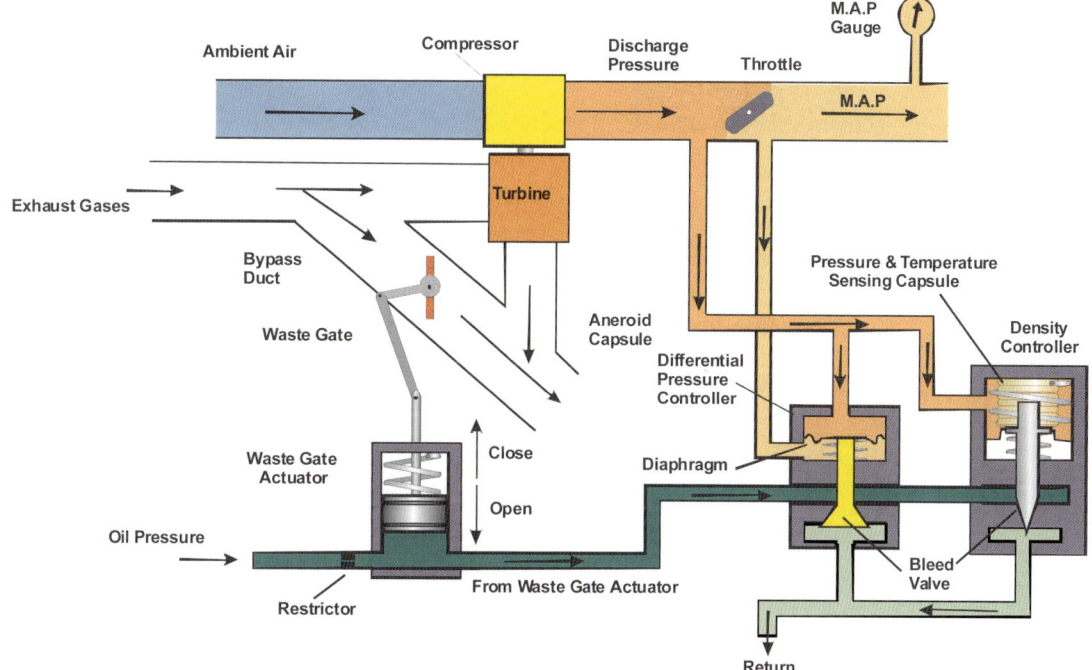

Fig. 9.12

These are fitted to some ground-boosted turbochargers and consist of density and differential pressure controllers. The density controller senses air density to prevent turbocharger outlet pressure from exceeding the maximum value and only regulates at full throttle up to the critical altitude. It has a capsule filled with dry nitrogen that is sensitive to both pressure and temperature. The capsule either expands or contracts in response to any change in pressure and temperature, varying the position of the spring-loaded bleed valve, therefore controlling the amount of oil bleeding from the waste gate actuator, resulting in repositioning of the waste gate. This ensures a constant density at full throttle.

The purpose of the differential pressure controller is to control the waste gate at all throttle positions other than fully open. It has a diaphragm dividing a chamber that has turbocharger outlet pressure on one side and manifold pressure on the other and responds to pressure drops across the throttle valve.

At full throttle, the bleed valve is fully closed with the least pressure drop. As the throttle closes, the bleed valve gradually opens, and the pressure drop increases. Thus, the controller opens the waste gate as the throttle closes, decreasing turbocharger outlet pressure in line with the selected power.

Slight changes in temperature or engine speed cause power variations resulting in a change in exhaust gas flow that affects turbine speed. An unstable condition of the manifold pressure, known as bootstrapping or hunting, may result as the control system attempts to reach a state of equilibrium. The differential pressure controller that quickly reacts to changes in the pressure drop across the throttle valve, reducing the effects of small power changes, counteracts this condition.

TRIPLE UNIT CONTROLLERS

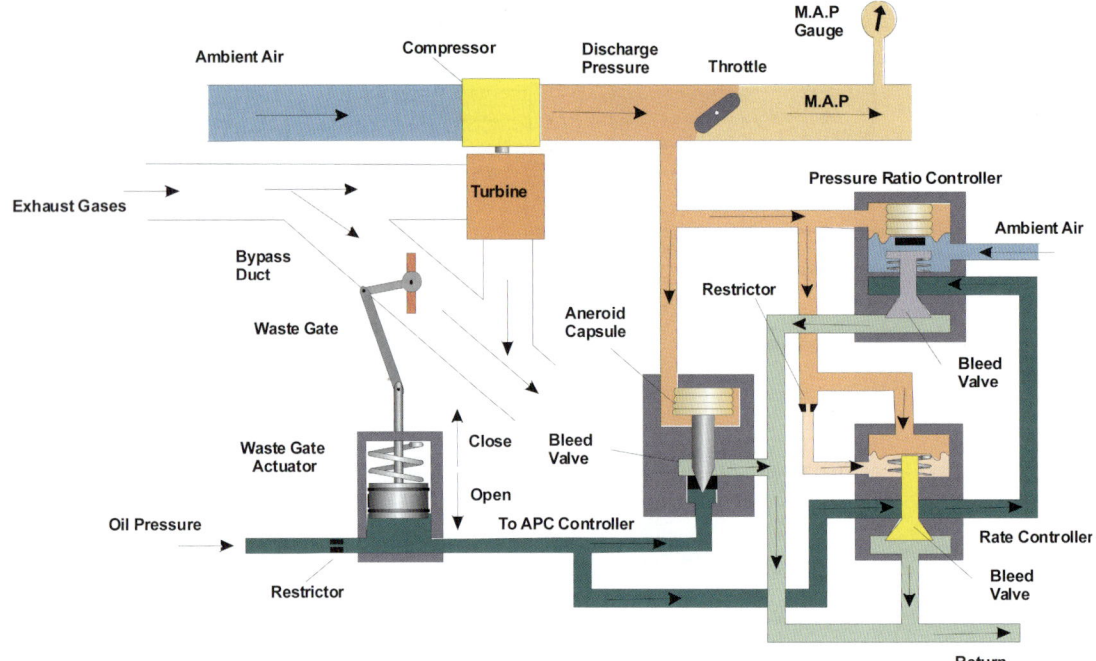

Fig. 9.13

These are also fitted to some ground-boosted turbochargers and consist of three separate controllers, two that control the waste gate up to critical altitude and the other to control the waste gate above critical altitude. Below critical altitude, an absolute pressure controller controls the turbocharger as previously described.

To control the rate at which turbocharger outlet pressure increases, a rate controller is fitted, which prevents the engine from overboosting initially when the throttle opens. A diaphragm in the unit is subjected to turbocharger outlet pressure on both sides. A restrictor is located in the opening to the lower chamber. Should turbocharger outlet pressure increase too quickly due to the restrictor, the air pressure above the diaphragm increases at a greater rate than it does below it. The resulting force on the diaphragm causes the spring-loaded bleed valve to open, thereby bleeding oil pressure from the actuator, opening the waste gate. In this way, the rate of outlet pressure increase remains controlled irrespective of the rate of engine acceleration.

With increasing altitude, the turbocharger must rotate faster to maintain maximum power by compressing more air. This results in increased temperature of the air supplied to the engine that could result in detonation. Therefore, placing limits on the maximum manifold pressure usable above a specific altitude controls this. Instead of retarding the throttle lever to operate within these limits above the specified altitude (typically 16 000 ft), a pressure ratio controller can be fit to limit turbocharger outlet pressure automatically above a specified altitude. It contains a chamber divided by a diaphragm with the upper chamber subjected to turbocharger outlet pressure and contains an aneroid capsule. The lower chamber is open to atmospheric pressure. As the aircraft climbs, the atmospheric pressure in the lower chamber progressively decreases allowing the aneroid capsule to expand. Upon reaching the specified altitude, the aneroid capsule in the chamber expands further and contacts the stem of the bleed valve. The bleed valve opens by an increasing amount as the aircraft climbs above the specified altitude, gradually increasing the bleed from the actuator and decreasing turbocharger outlet pressure at a set ratio to the atmospheric pressure of usually 2.2-to-1.

INTERCOOLER

Due to compression, the air temperature increases as it flows through the compressor of a turbocharger or supercharger. An intercooler is installed in some systems to cool the air after compression, therefore increasing its density without decreasing its pressure. In some installations, intercooler shutters may be fitted in order to control cooling air.

TURBO LAG

Since the turbocharger rotates at a very high speed, it needs time to accelerate and stabilise, therefore requiring smooth and slow throttle movements. This is known as **turbo lag.**

Chapter 10
Propellers

INTRODUCTION

This chapter does not cover propeller terminology or aerodynamics of the propeller, as they are covered in Principles of Flight. A propeller is a means of converting engine power into propulsive force. Rotating a propeller results in the rearward acceleration of a mass of air, the reaction to this rearward motion is a forward force on the propeller blades called **thrust**, where:

$$\text{THRUST = MASS AIRFLOW x ACCELERATION}$$

The propeller accelerates a large mass of air rearward at a relatively low velocity. The reaction to this is thrust force acting in a forward direction, propelling an aircraft along its flight path. When a propeller is fitted in front of an engine, it is a **tractor**, whereas when fitted at the rear it is a **pusher**.

PROPELLER EFFICIENCY

The definition of propeller efficiency is the ratio of **thrust horsepower** (THP), which is delivered by the propeller, to the engine power (BHP) required to drive the propeller at a given rpm, expressed as a percentage. Thus:

$$\text{PROPELLER EFFICIENCY} = \frac{\text{THRUST HORSEPOWER}}{\text{BRAKE HORSEPOWER}} \times \frac{100}{1}$$

Another definition is the ratio of useful work done by the propeller in moving an aircraft, to the work supplied by the engine. The work done by the propeller is the product of the thrust and forward airspeed (TAS).

The work supplied by the engine is the torque required to turn the propeller at a given rpm. Thus:

$$\text{PROPELLER EFFICIENCY} = \frac{\text{THRUST X TAS}}{\text{PROPELLER TORQUE X rpm}}$$

When the aircraft is stationary on the ground with the engine running, the propeller is 0% effective, since, although it may be developing a lot of thrust, it is not doing any work. As the forward speed of the aircraft increases, the efficiency increases. Expect a level of up to 88% upon achieving the optimum airspeed for that propeller.

FIXED PITCH PROPELLERS

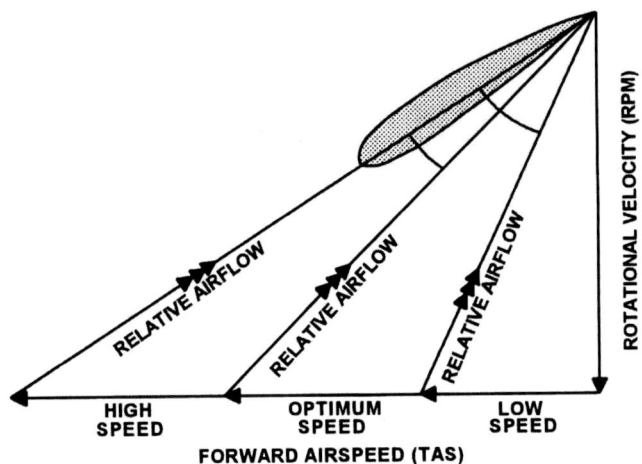

Fig. 10.1

Figure 10.1 illustrates a fixed pitch propeller travelling at different forward speeds and at a constant rpm. For a fixed blade angle, with variations in forward speed the angle of attack changes. As the forward speed increases, the angle of attack decreases and with it thrust.

FIXED PITCH PROPELLER DISADVANTAGES

Fixed pitch propellers, like most aerofoils, are most efficient only under one set of conditions. That condition is typically cruise, but until reaching cruise airspeed, the angle of attack of the propeller blades is comparatively large, therefore the propeller is less efficient.

During take-off, the angle of attack of the blades of such a propeller would be extremely large, and result in poor acceleration, hence requiring a longer take-off run. In the cruise condition, the angle of attack is at its optimum (small angle), therefore limiting forward speed to prevent engine overspeed. When a fixed pitch propeller is optimised for take-off and climb performance, the cruise speed of the aircraft is compromised, since the blades' angle of attack would be too low for maximum efficiency at higher speeds.

These disadvantages led to the development of variable pitch or constant speed propellers. Figure 10.2 illustrates the characteristics of various fixed pitch propeller angles versus a constant speed propeller at various airspeeds.

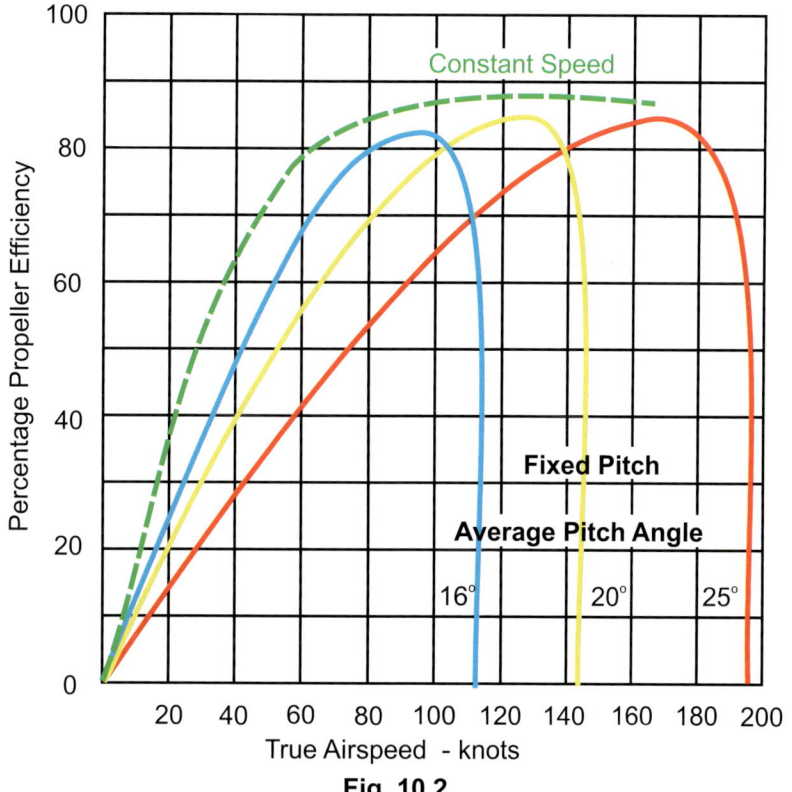

Fig. 10.2

For the propeller to be efficient over the whole operating range, the blade angle needs to vary to maintain the optimum angle of attack of the blade, which is approximately 2° to 4°. As forward speed increases, the blade angle must increase to maintain the same angle of attack, as illustrated in figure 10.3.

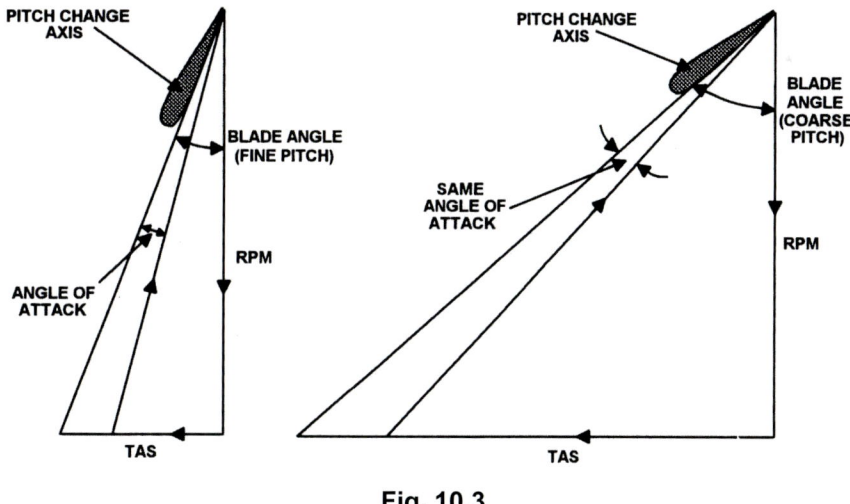

Fig. 10.3

BLADE TWIST

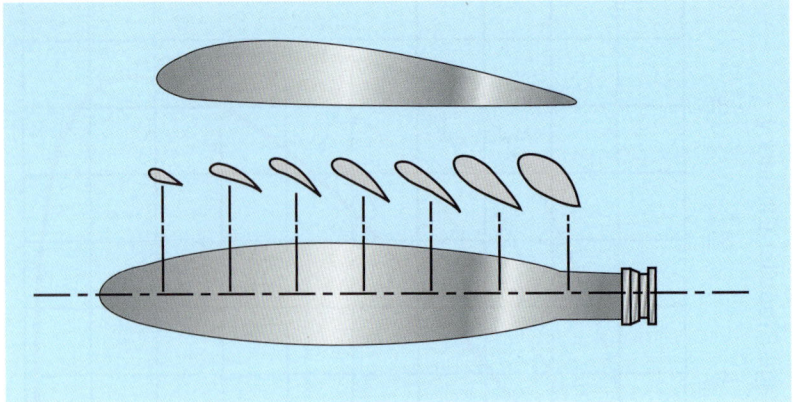

Fig. 10.4

Since each blade is an aerofoil cross section, it produces thrust most efficiently at a particular angle of attack. This angle varies both with operating conditions and with the design camber of the blade sections. The rotational speed of a particular cross section of a blade increases with its distance from the axis of rotation, and since the forward speed of all parts of the blades is the same, the relative airflow varies along the blade. It is therefore necessary to provide a decreasing blade angle from root to tip.

THE VARIABLE AND CONSTANT SPEED PROPELLER

A variable pitch propeller allows the blade angle to vary in flight in order to fully utilise engine power. The original variable pitch propeller had two blade settings: a fine pitch for take-off and climb, and a coarse pitch to enable full engine speed for use in cruising.

The introduction of an engine-driven propeller governor enables the blade angle to alter automatically, defining it as a constant speed propeller. The pitch setting varies automatically to maintain a pre-selected constant rotational speed. As a result, the engine and the propeller can work at their maximum efficiency, regardless of whether the aircraft is at take-off, climb, cruise, or maximum speed. The blade pitch varies by the operation of the propeller governor that controls oil flow in and out of a propeller pitch change mechanism to move a piston. The piston connects to the propeller blade, thereby changing the pitch angle.

The governor has two names, **constant speed unit (CSU)** in the case of a piston engine, or **propeller control unit (PCU)** for a turboprop. The operation of these is described later. The term variable pitch is for use when describing a constant speed propeller.

CONSTANT SPEED PROPELLER BLADE POSITIONS

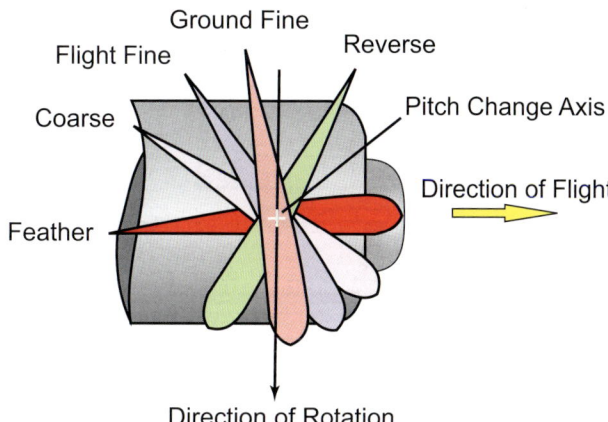

Fig. 10.5

There are various blade pitch positions associated with different constant speed propeller assemblies as shown in figure 10.5 and described below:

> **Feathered**
>
> When the chord line of the blade is parallel to the airflow, therefore preventing windmilling.

> **Coarse Pitch**
>
> The maximum cruising pitch in normal operation.

> **Flight Fine Pitch**
>
> The minimum pitch obtainable in flight.

> **Ground Fine Pitch**
>
> The minimum torque position for ground operation and is sometimes referred to as **superfine pitch**.

> **Reverse Pitch**
>
> An aerodynamic brake position used for braking and sometimes ground manoeuvring. It is achieved by accelerating air forward by the blade going into a negative angle.

> **Alpha Range**
>
> The flight operating range, from flight fine pitch to coarse pitch.

> **Beta Range**
>
> From flight fine pitch to reverse pitch which is the ground operating range and is hydro-mechanically controlled by a flight deck power lever.

SINGLE-ACTING PROPELLER

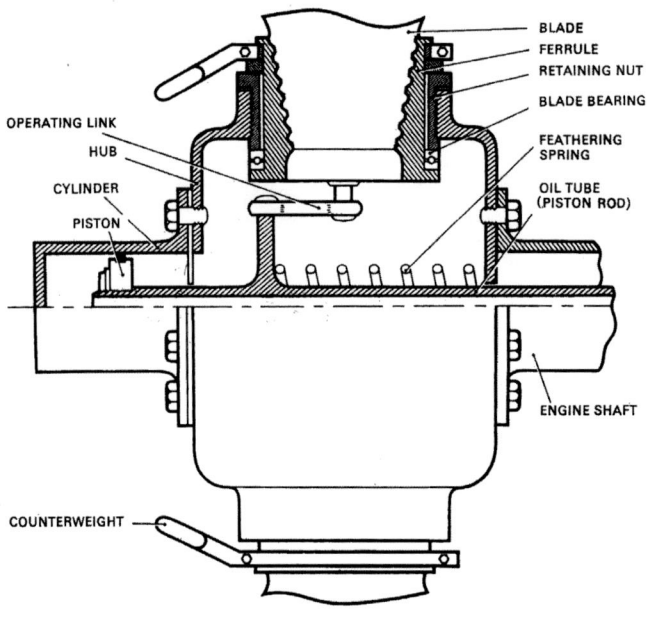

Fig. 10.6

This is the type of propeller normally fitted to a light piston engine aircraft. The pitch change mechanism consists of a piston housed in a cylinder. The piston connects to the propeller blade via an operating link. One side of the piston is subject to boosted engine oil pressure whilst the other side is subject to spring force.

On a constant speed feathering propeller, fitted onto a light twin-engine piston aircraft, the boosted oil pressure plus blade **centrifugal turning moment (CTM)** turns the propeller to fine pitch. Movement to coarse pitch and feather is achieved via the spring and counterweights attached to the blades once the oil pressure has been relieved through the constant speed unit.

On a single engine non-feathering propeller there are no counterweights, so boosted oil pressure is used to turn the blades to coarse, and blade CTM and a light spring turn the blades to fine.

LOW PITCH STOP OR CENTRIFUGAL LATCH

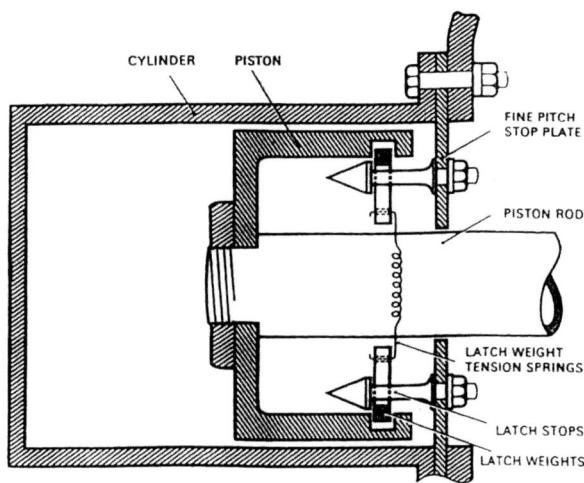

Fig. 10.7

Sometimes referred to as a start latch, it is fitted to prevent the propeller from turning to feather when the engine shuts down. The blades normally feather due to the oil pressure bleeding away through the CSU, allowing the spring force to turn the blades to the feather position. If the blades feather it places an unacceptable load on the engine during start. A centrifugal latching mechanism locks the piston in fine pitch. Once the engine starts and the rpm increases above ground idle the centrifugal force removes the latches, allowing piston movement. Decreasing the rpm to the ground idle value results in the return springs engaging the latches, locking the blades in the fine pitch position.

It is important not to shut down the engine from a high rpm on the ground, otherwise the latches do not engage and the propeller feathers. In addition, take care in flight to prevent the rpm from falling such as due to engine failure below a pre-determined figure, usually between **800 to 1000 rpm**, as the latches engage and prevent feathering. Refer to the current AIC for more information.

CONSTANT SPEED UNIT (CSU)

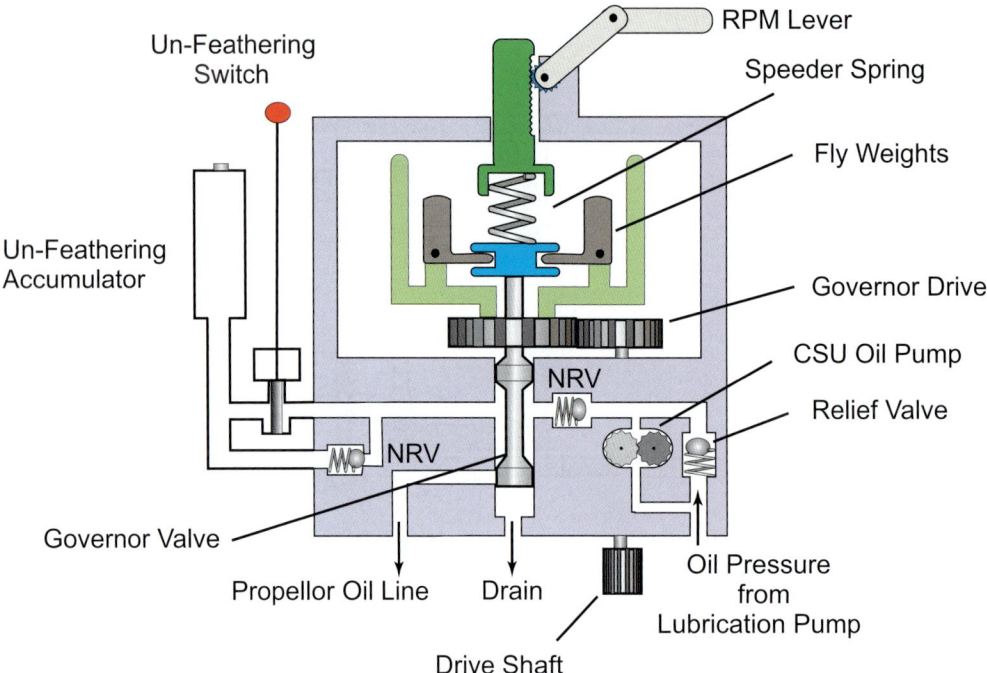

Fig. 10.8

The **constant speed unit (CSU)** of a single-acting propeller controls the oil pressure to the piston to move the blades to fine pitch. To move the blades to coarse pitch, relieving the oil pressure causes the spring to move the piston to coarse. The CSU consists of centrifugal flyweights, a control valve, a control spring (speeder spring), a non-return valve, and an oil pump to boost engine oil pressure for propeller control mechanism operation.

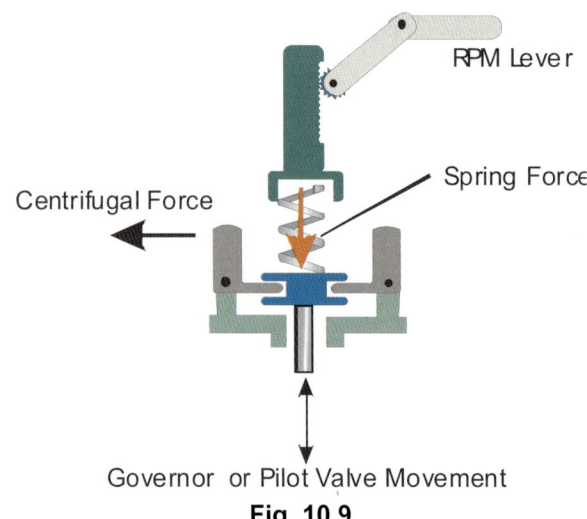

Fig. 10.9

Driven by the engine, the governor's L-shaped flyweights measure the engine speed. If engine speed increases, the flyweights move outward under centrifugal force, lifting the control valve against the opposing control spring. The rpm lever sets the control spring tension. The engine speed and spring force determines the control valve position. With these forces in balance, the control valve is in the closed position, preventing oil from flowing into or from the cylinder and creating a hydraulic lock, which prevents movement of the piston. The principle of operation is also applicable to double-acting propellers.

The rpm lever that controls the tension of the control spring achieves the desired rpm selection. Moving the rpm lever fully forward, with the throttle at a low power setting, to the maximum rpm position fully compresses the control spring and oil from the pump goes to the propeller operating mechanism turning the propeller to fully fine. At the low power setting, any increase in throttle position results in the engine reacting in the same manner as a fixed pitch propeller until reaching the CSU speed range.

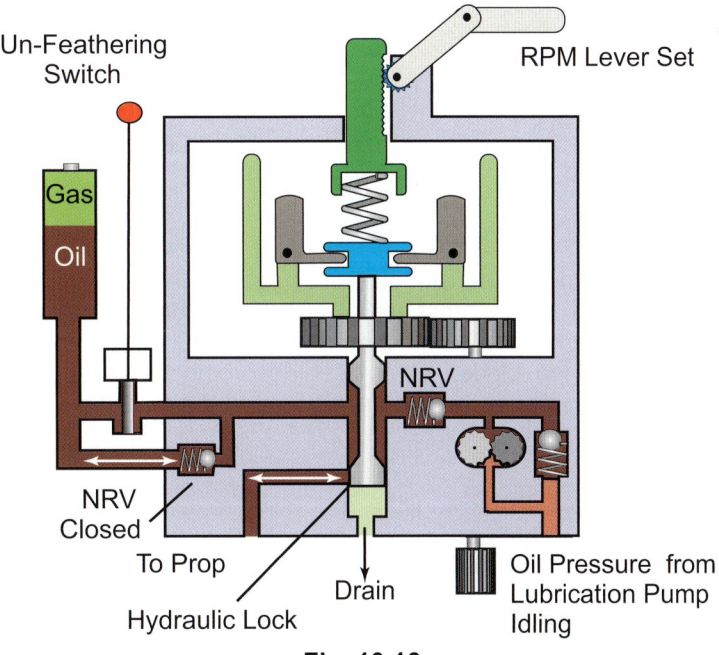

Fig. 10.10

After this point, if the throttle opens to increase power and engine speed, the centrifugal force on the flyweights raises the control valve until it reaches the position where it obtains maximum rpm. The centrifugal force of the flyweights and spring control force are in balance and the CSU is in the on-speed condition, as shown in figure 10.10. As a result, there is no oil flow in or out of the CSU.

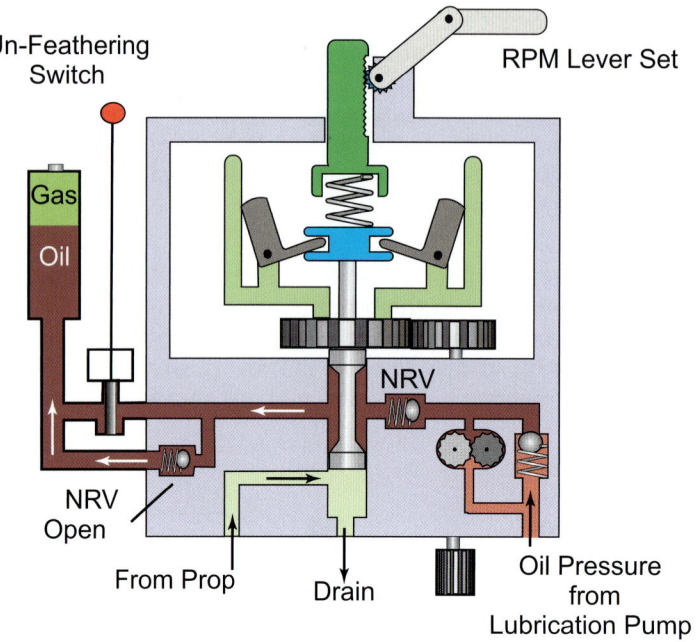

Fig. 10. 11

Should the propeller load decrease the rpm increases above the rpm lever setting. The flyweight centrifugal force exceeds the spring pressure causing the pilot valve to rise, thus moving the governor to an **overspeed condition** as illustrated in figure 10.11. Oil drains from the propeller causing the piston to move the blades to coarse, absorbing the reduced propeller load and preventing the rpm increasing above its pre-determined value with the CSU assuming the on-speed condition once more.

When selecting a reduced rpm by moving the rpm lever rearward, the speeder spring tension reduces and results in the flyweights raising the control valve, which allows the oil to drain from the propeller hub coarsening the blades. The increased engine load decreases engine rpm until the centrifugal force of the flyweights balances the speeder spring force at the new selected rpm, and the CSU assumes the on-speed condition.

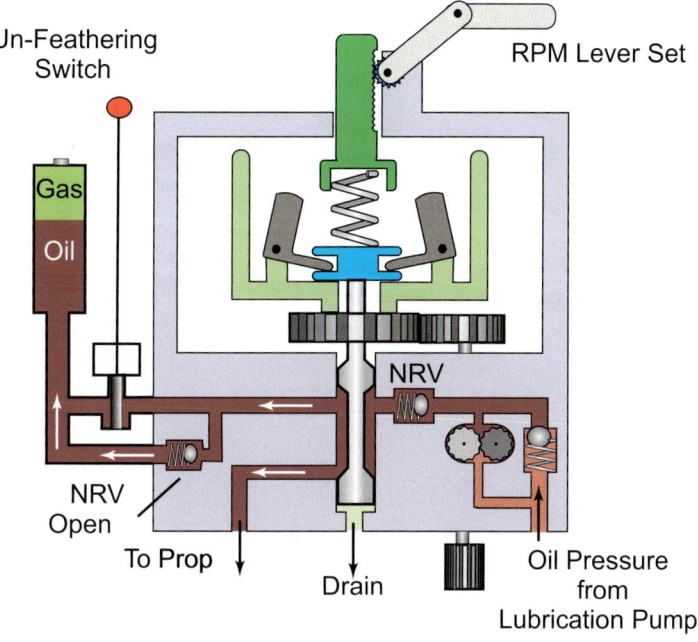

Fig. 10.12

Alternatively, increased propeller load decreases the engine rpm, allowing the speeder spring force to overcome the centrifugal force of the flyweights. The flyweights move inward, causing the control valve to drop. The CSU is now in an **underspeed condition,** as shown in figure 10.12. Oil now goes into the propeller hub and the blades move toward fine pitch, decreasing the propeller load, maintaining the pre-selected rpm. Once again the CSU assumes the on-speed condition.

Moving the rpm lever forward commands an increase in rpm, therefore increasing the speeder spring tension and causing the flyweights to move inward. This allows the control valve to drop and direct oil into the propeller hub, decreasing the blade angle. This results in the rpm increasing until achievement of the on-speed condition at the selected rpm. In basic terms, if the control valve lifts, the blades go toward coarse and ultimately feather. If the control drops, the blades go toward fine. This is true as long as the operation is within the controlling speed range of the CSU.

SINGLE-ACTING PROPELLER FEATHERING

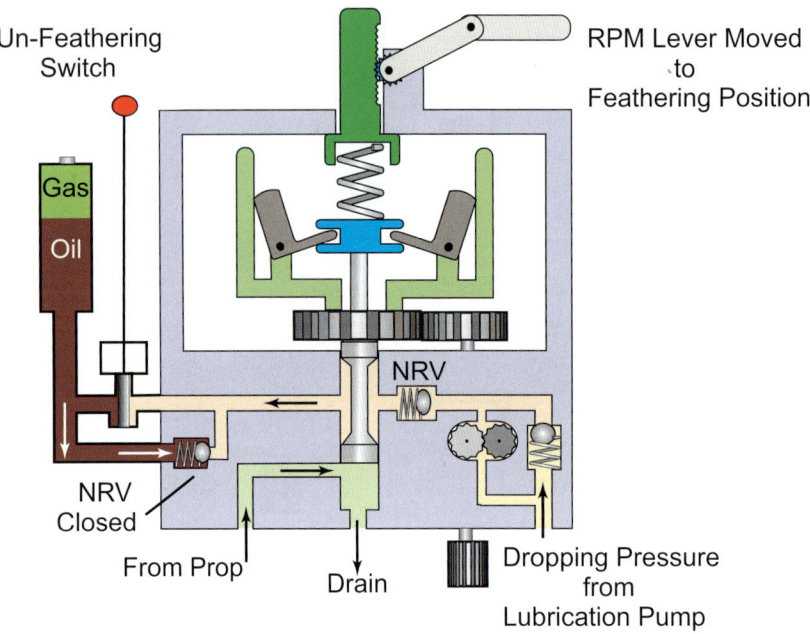

Fig. 10.13

Feathering is a procedure of turning the propeller blades until the blade chord line is almost, or completely parallel to the airflow, so there are no rotating forces acting on the propeller that produce high drag due to windmilling. This is desirable to prevent further rotation of the engine that may cause damage and ensures minimum drag is maintained. In order to achieve feather, the control valve must lift to release the oil pressure, allowing the feathering spring and counterweights to move the blades to feather, as in figure 10.13.

Moving the rpm lever fully rearward past the low rpm position to the feather position releases the control spring pressure and mechanically lifts the control valve up. An internal feather stop is fitted to limit piston travel. Once in the feather position, the rpm lever, depending on the installation, may be locked in position by a feather latch to prevent inadvertent de-selection.

SINGLE-ACTING PROPELLER UN-FEATHERING

When feathered, there is little or no propeller torque to turn the engine for restart, if desired. To un-feather a propeller, oil pressure is necessary to turn the propeller from the feathered position toward coarse in order to create windmilling to restart the engine. Two methods can supply oil pressure, either by cranking the engine on the starter or by use of an un-feathering accumulator that stores oil pressure. The accumulator is a cylinder that contains a separator piston subject to CSU oil pressure on one side and nitrogen pressure on the other. The accumulator charges with oil pressure when the engine is running and a non-return valve retains the pressure.

When using the engine cranking method, place the rpm lever in the correct position, which is normally just forward of the minimum rpm position or just out of the feather gate, in order to prevent engine overspeed. This allows the CSU control valve to drop under action of the control spring. This admits oil pressure created due to the starter turning the engine to the pitch change mechanism and moving the propeller from the feather position. The disadvantage of this system is the high loads placed on the starter motor due to the high torque reaction of the propeller blades in the feathered position.

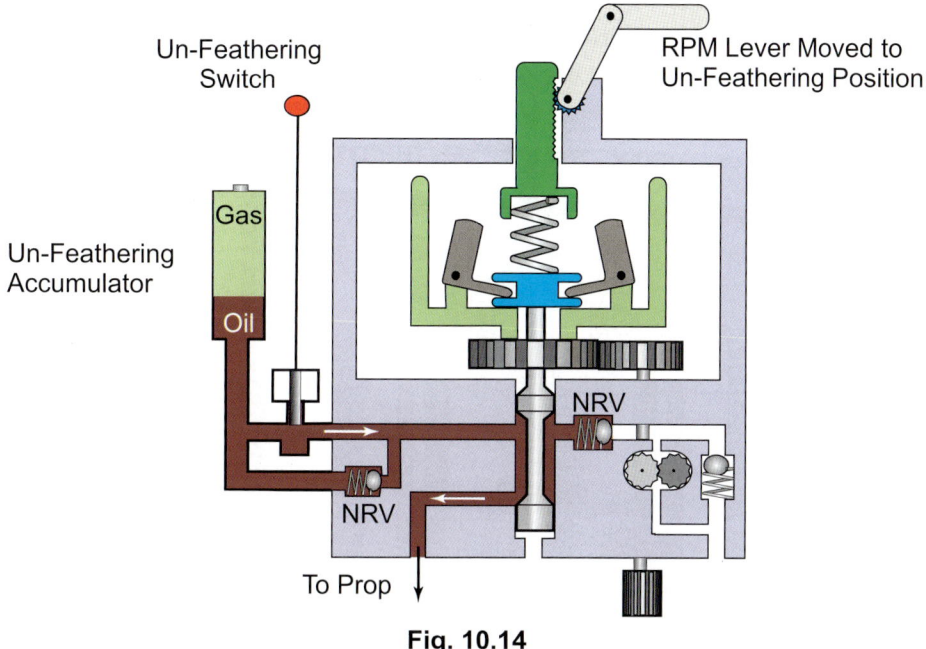

Fig. 10.14

The fitting of the un-feathering accumulator overcomes the disadvantage of the cranking method. As before, moving the rpm lever forward to the correct position for un-feather allows the CSU control valve to lower. In this case, the accumulator supplies the oil pressure necessary by releasing the stored oil pressure to the pitch change mechanism. Depending on the system employed, oil pressure may be released either by forward movement of the rpm lever only or in conjunction with the operation of an un-feathering button. The propeller moves from feather toward coarse causing windmilling of the propeller that in turn rotates the engine and effects engine start without the aid of the starter motor.

PROPELLER CONTROL UNIT (PCU)

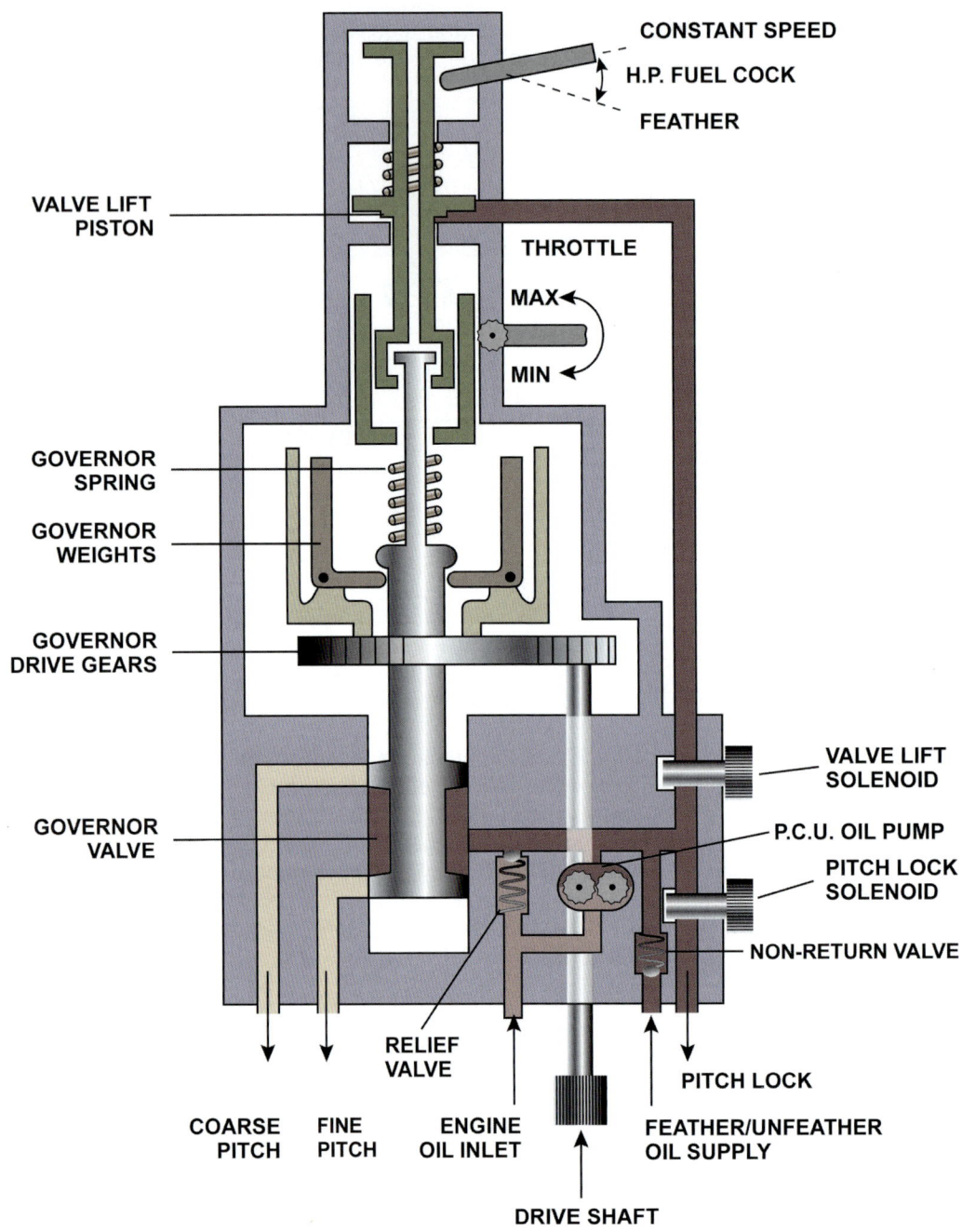

Fig. 10.15

These are fitted to a turbo-propeller engine and function in the same manner as a CSU but incorporate various extra functions. In this system, it is common for the engine fuel control unit and PCUs to connect to the power control lever. As a result, fuel flow and engine speed are selected together. Other installations can incorporate a separate rpm control or condition lever. Figure 10.15 illustrates a typical PCU.

DOUBLE-ACTING PROPELLER

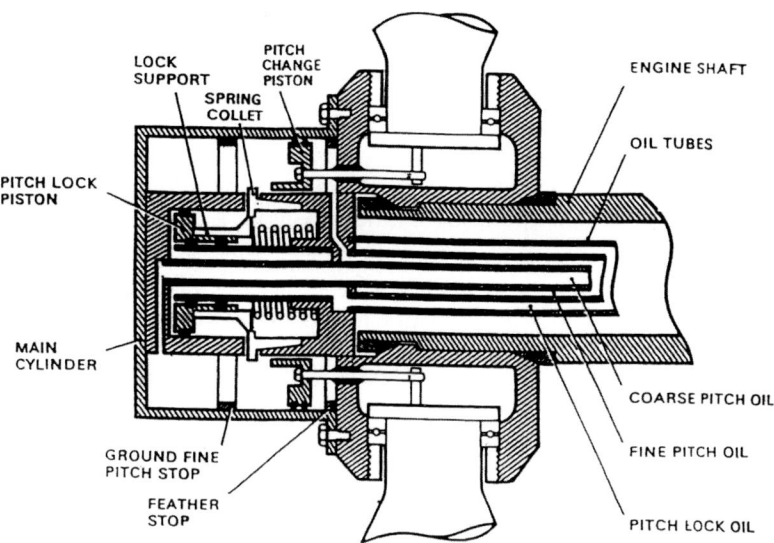

Fig. 10.16

The double-acting propeller is the type usually fitted to larger engines. In this arrangement, there are no counterweights fitted and the piston is subject to oil pressure on both sides, turning the propeller to fine, coarse, and feather. In the case of a turboprop, a PCU controls the flow of oil.

DOUBLE-ACTING PROPELLER FEATHERING

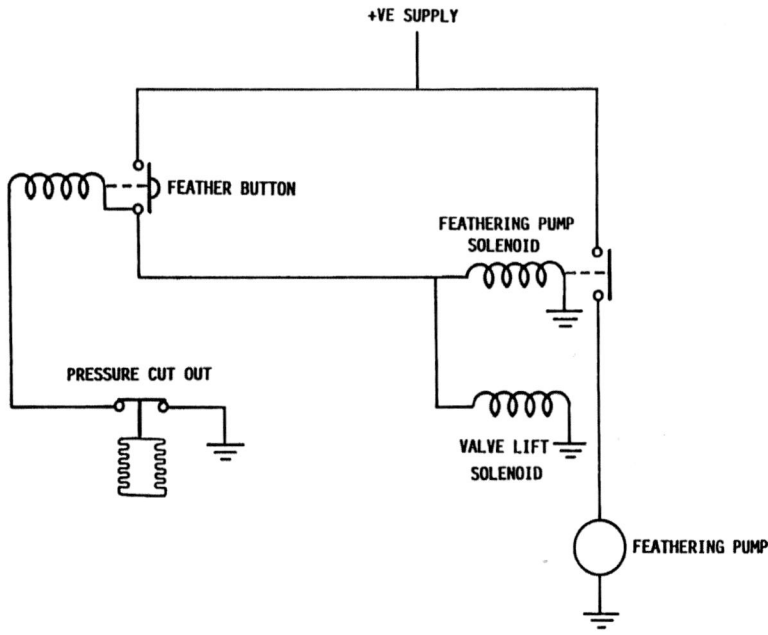

Fig. 10.17

Since a double-acting propeller operates by directing oil pressure to either side of the piston in the pitch change mechanism, oil pressure is required in order to feather. Fitting an electrical oil pump in the system that takes oil from the bottom of the oil tank below a stack pipe achieves this. Figure 10.17 shows a simplified typical feathering circuit.

Pushing in the feathering button (normally illuminated) energises a holding coil. This activates the electrical pump to supply oil pressure. It also energises a valve lift solenoid, allowing the pump oil pressure to lift the control valve, allowing pump oil pressure into the pitch change mechanism to feather the propeller. Once reaching the full feather position, a pressure cut out switch turns off the feathering pump.

On a manual system, moving the high pressure fuel cock to the feather position, mechanically lifting the control valve, lifts the control valve in the PCU. If insufficient oil pressure is available from the engine-driven PCU pump to move the propeller to feather, then operation of the electrical feather pump becomes necessary.

DOUBLE-ACTING PROPELLER UN-FEATHERING

In a double-acting propeller, the electrical feathering pump oil pressure directed to the pitch change mechanism achieves un-feathering, with the power levers closed and the high pressure fuel cock open. The rpm lever moves in normal operating range and the control valve lowers under the action of the governor spring. The electrical feathering pump switches on and oil pressure discharges to the PCU, turning the propeller from feather toward coarse. The propeller then windmills and rotates the engine. Once the engine starts and is on speed, the oil pressure from the feathering pump rises and a cut out switch turns the pump off. Operation of the pump occurs either via manual selection of a switch or automatically via a micro switch mounted on the high pressure fuel cock lever.

PITCH STOPS

These are fitted to control the propeller angle for ground and flight operations. The types of pitch stops are:

> **Ground Fine Pitch Stop**
> This stop type ensures fine pitch on the ground during engine start and ground running.

> **Flight Fine Pitch Stop**
> This stop type limits the minimum pitch in flight to prevent overspeed and resulting high drag. It must be removed to allow selection of ground fine pitch for ground operation.

BETA RANGE

Some gas turbine engines use a form of control known as Beta Control. Beta is blade angle, and during ground operations only, direct control of the propeller pitch by the power levers is achieved in the ground idle and reverse pitch range. To operate in the beta range, the aircraft must be on the ground and have the flight fine pitch stops removed. This gives better control for ground manoeuvring.

REVERSE PITCH

In ground fine pitch, the blade position is 0° and provides high windmilling drag to aid aircraft retardation on the ground to a low forward speed. To improve this on slippery or short runways, some engine installations are fitted with reverse pitch propellers. This system includes installation of removable ground fine pitch stops. With the ground fine pitch stop removed and reverse selected, moving the power levers rearward beyond ground idle causes the blades to move to a negative pitch, applying the correct amount of engine power to produce reverse thrust.

PITCH LOCKS

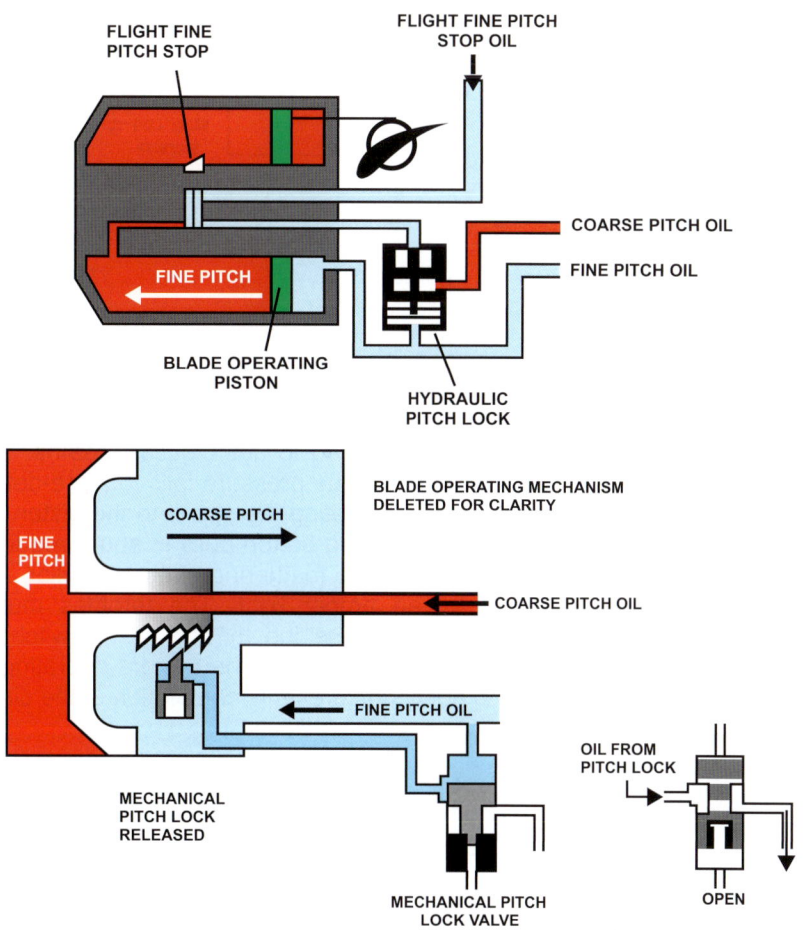

Fig. 10.18

Pitch locks lock the blades at whatever angle they are currently at should there be a propeller mechanism or PCU failure, which would cause the propeller to run to fine due to CTM. There are various types of lock, two of which appear above:

- ➢ **Hydraulic Lock**

  This responds to fine pitch oil pressure failure to create a hydraulic lock.

- ➢ **Mechanical Lock**

  Again, this responds to fine pitch oil pressure failure and mechanically locks the blade.

AUTOMATIC FEATHERING

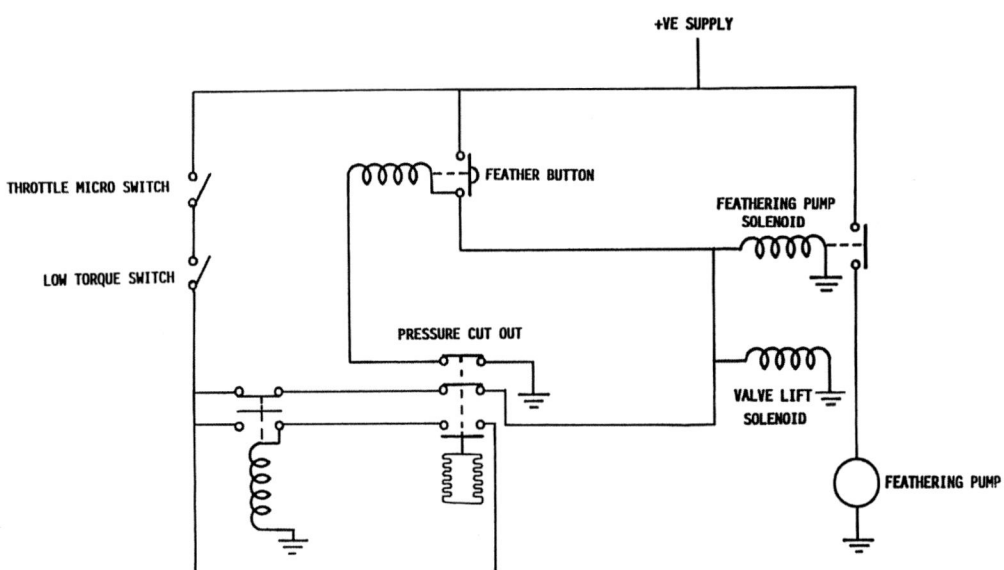

Fig. 10.19

An automatic feathering system is sometimes provided to automatically feather the propeller in the event that engine power and hence indicated torque pressure falls to a pre-determined value. In this instance, a low torque switch operates, completing the circuit to the piston lift solenoid on the PCU and feathering pump. The relevant feathering button pulls in and a red light illuminates. The control valve rises hydraulically, thus enabling the feathering of the propeller. A switch on the flight deck arms the system, indicated by an amber light. The throttles must advance to approximately 45 to 75% of lever movement to close the throttle micro switch. Normally this system is only used during take-off and landing. To prevent the system operating as a result of momentary loss of torque pressure, a time delay unit prevents completion of the circuit until a pre-determined time has elapsed, typically one or two seconds.

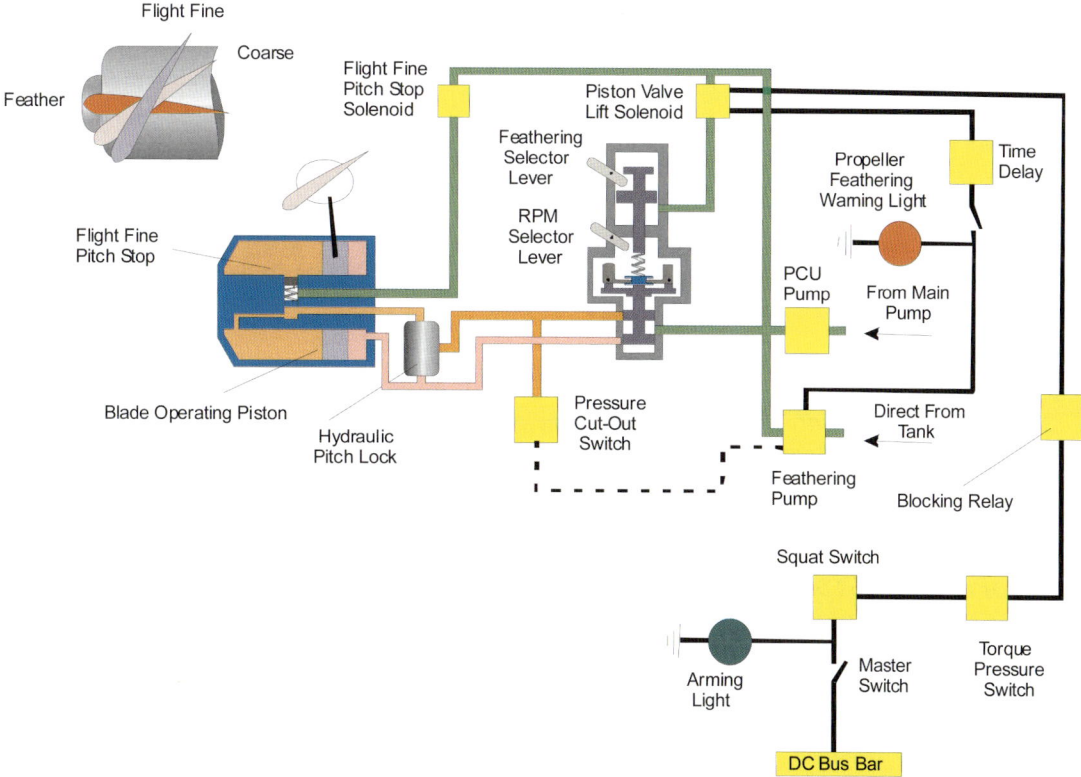

Fig. 10.20

To prevent more than one engine from autofeathering, a blocking relay is usually fitted either between the master switch and the throttle switch, or incorporated in the feathering button circuit. Sometimes it can be reset to re-arm the autofeather system in the event of another engine failure. By activating the feather button, regardless of whether or not the propeller has been auto-feathered, any engine can be feathered at any time.

Some engines incorporate an **automatic drag limiting (ADL)** system or **negative torque sensing (NTS)** system that do not feather the propeller in the event of engine failure but turn the blades to coarse to limit windmilling.

SYNCHRONISATION SYSTEM

The purpose of a synchronisation system is to reduce vibration and cabin noise by ensuring that all engines are set to the same rpm. One engine is the master engine, whilst the other engine(s) is the slave engine(s). In the case of four-engine aircraft, any engine can serve as the master, but the master is always the left engine on a light twin-engine aircraft. Comparison of electrical signals generated from the engines occurs and if an imbalance exists between the rpms, then the slave engine(s) governor automatically adjusts to match the master engine rpm. For the system to operate, the slave engine rpm must be within a certain speed of the master engine. A typical value is 100 rpm. This system is not for use during take-off or landing as failure of the master engine would result in a tendency for the slave engine(s) to follow the master resulting in a loss of power.

SYNCHROPHASING SYSTEM

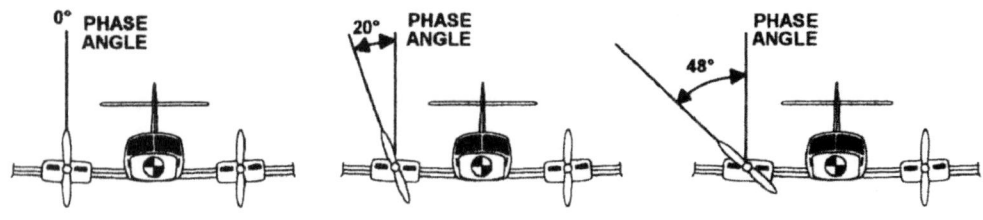

Fig. 10.21

To further improve vibration and noise reduction, a synchrophaser system is used. It involves phasing the propeller relative positions at any specific time and enables the blades of the slave engine(s) to be set a number of degrees in rotation behind that of the master engine. Most systems both synchronise and synchrophase at the same time.

SYNCHROPHASING SYSTEM OPERATION

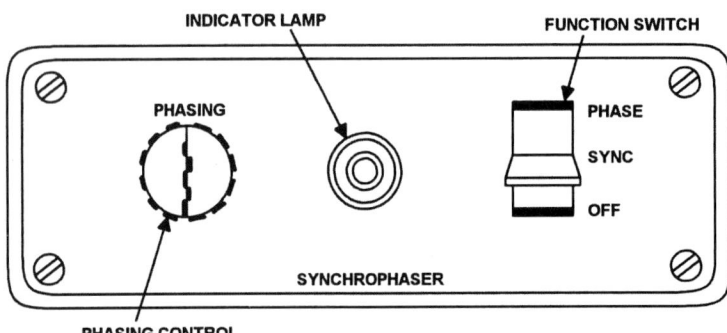

Fig. 10.22

A typical system consists of magnetic pickups on each propeller, trimming coils on the propeller governor and a control box. The magnetic pickups send speed and phase angle information from all engines to the control box. The control box compares the signals and sends a signal to the propeller governor(s) trimming coils, which adjusts the appropriate phase angle whilst maintaining the pre-selected rpm. In larger aircraft, a flight deck propeller phase control is for use in selecting the phase angle that provides minimum vibration. On a light aircraft system, a switch that only allows a choice of two pre-set phase angles may select synchrophasing.

As before, the engines must be within a certain speed range before the system is selectable. The speed range can be as low as 10 rpm in the case of a light aircraft up to typically 100 rpm on larger aircraft. The indicator lamp flashes the entire time the engines are out of synchronisation, extinguishing when they are in sync.

PROPELLER CHECKS

Propeller checks ensure the propeller governor and operating mechanism are functioning correctly. The rpm lever should be positioned fully forward in the maximum rpm position; the throttle is then set to the rpm setting for the engine. The rpm lever is moved from MAX to MIN, observing the drop in rpm (on some engines move the rpm lever until the rpm drops by a specified amount). The rpm lever is then returned to the MAX position, ensuring the restoration of the original rpm figure. Carry out this exercise three times to ensure that the propeller operating mechanism is charged with low viscosity hot oil thereby preventing sluggish operation.

REDUCTION GEARING

The purpose of reduction gearing is to enable the propeller to rotate at the most efficient speed to absorb the engine power, whilst the engine rotates at a higher speed to develop more power. This is particularly the case when operating a turbo-prop. Reduction ratios can vary from 2:1 to 15:1 depending on the power unit employed. Typical examples of gearing design are:

> ➤ Spur gear
> ➤ Planetary gears
> ➤ Bevel planetary gears
> ➤ Combination of spur and planetary gears

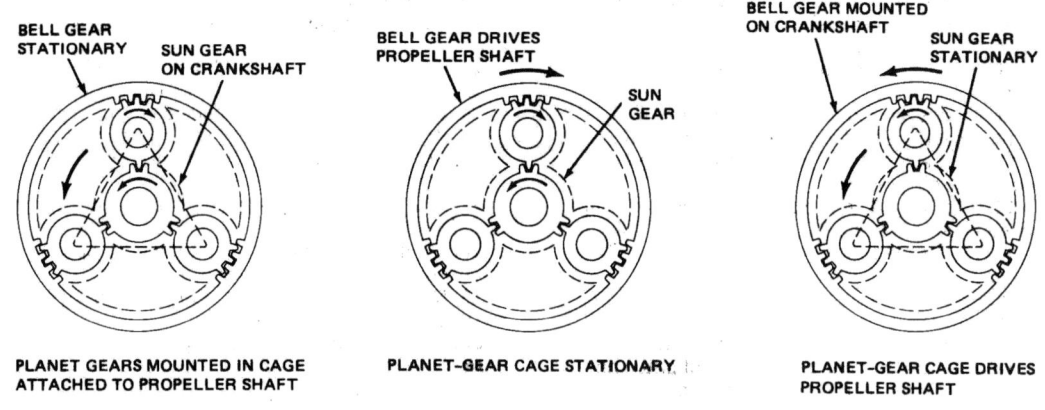

Fig. 10.23

Figure 10.23 illustrates typical planetary gear arrangements.

TORQUEMETERS

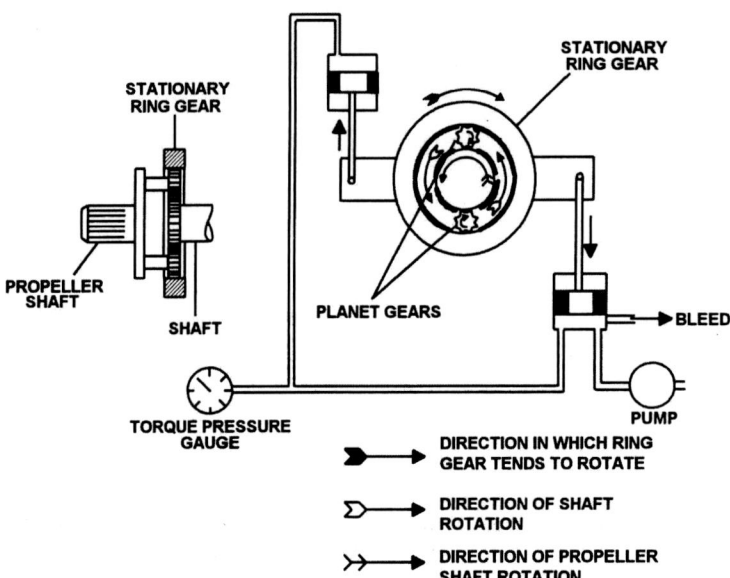

Fig. 10.24

Power produced by a propeller is proportional to the torque, where torque is the turning moment that is produced by the propeller around the axis of the output shaft. A torquemeter on the flight deck indicates the power produced by a turbo-propeller engine. There are various torquemeter systems; a typical system appears in figure 10.24. It is part of the engine, normally assembled within the reduction gear assembly between the output and propeller shafts.

System operation is based on the principle of the tendency for part of the reduction gear to rotate, which is resisted by hydraulic cylinder pistons. Pressure created by the pistons transmits to a flight deck gauge that can display as pressure in pounds per square inch (psi) or shaft horsepower. The greater the pressure indication the greater the torque, and therefore power, and vice versa.

Torque measurement can also occur via an electrical strain gauge system consisting of a fine insulated conductor wire bonded to a component and consisting of two independent bridge circuits. Upon applying strain, a transducer generates a millivolt electronic signal proportional to engine torque. A signal conditioner amplifies the input signal and provides a varying voltage to the flight deck indicator, which is essentially a voltmeter that may display engine power as psi, horsepower, or percent power or percent torque.

Chapter 11
Gas Turbine
Principles of Operation

INTRODUCTION

The function of any propeller or gas turbine engine is to produce a propulsive force, known as thrust, by accelerating a mass of air or gas rearward. Therefore, knowledge of Newton's Laws of Motion greatly simplifies the understanding of the production of thrust.

NEWTON'S LAWS OF MOTION

The three laws of motion are:

1st Law states: That a body will continue in a state of rest, or uniform motion in a straight line, unless acted upon by an external force.

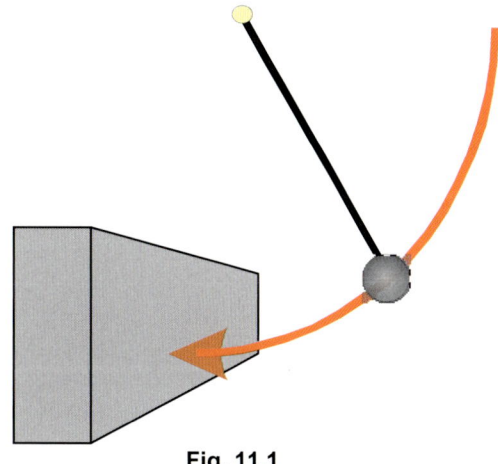

Fig. 11.1

2nd Law states: That a body at rest or in uniform motion will, when acted upon by an external force, accelerate in the direction of the force. The magnitude of the acceleration for any given mass is directly proportional to the size of the applied force.

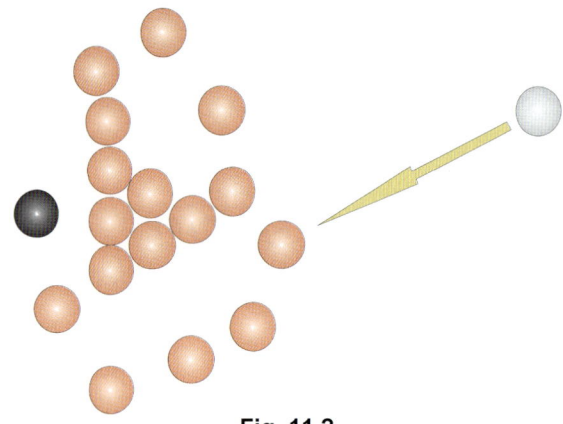

Fig. 11.2

3rd Law states: For every action there is an equal and opposite reaction.

Thrust   Gas Mass Acceleration

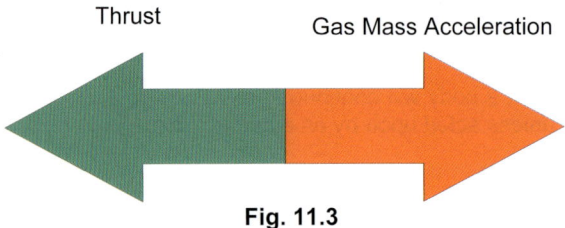

Fig. 11.3

The third law is most applicable to the operation of gas turbine engines since in the operation of a gas turbine engine the gas mass accelerates in a rearward direction, thus, by reaction, producing thrust.

BERNOULLI'S THEOREM

Point 1   Point 2   Point 3

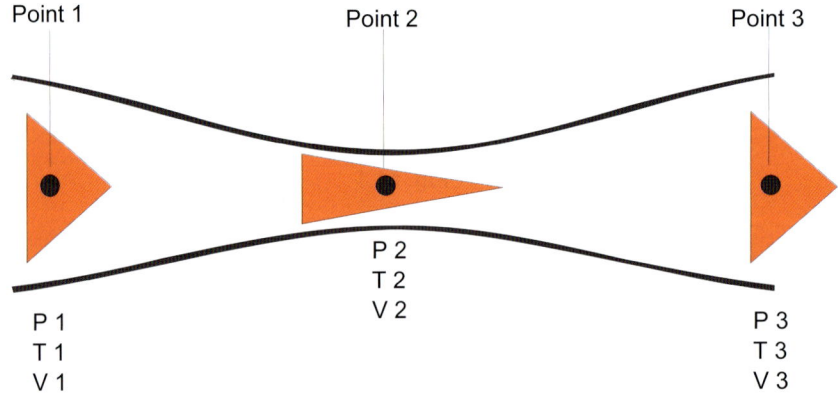

P 2
T 2
V 2

P 1   P 3
T 1   T 3
V 1   V 3

Total Energy at Points 1, 2, and 3 are equal. To maintain these values, Pressure, Volume, and Temperature must alter.

Fig. 11.4

At any point in a tube (or a gas passage) through which liquid (or gas) is flowing, the sum of the pressure energy, the potential energy, and the kinetic energy remains constant. Thus, if one of the energy factors in a gas flow changes, one or both of the other variables also change, so that the total energy remains constant.

This theorem gives us the relationship between velocity and pressure of a stream of air flowing through a tube, or duct, such as a gas turbine engine.

CONVERGENT DUCT

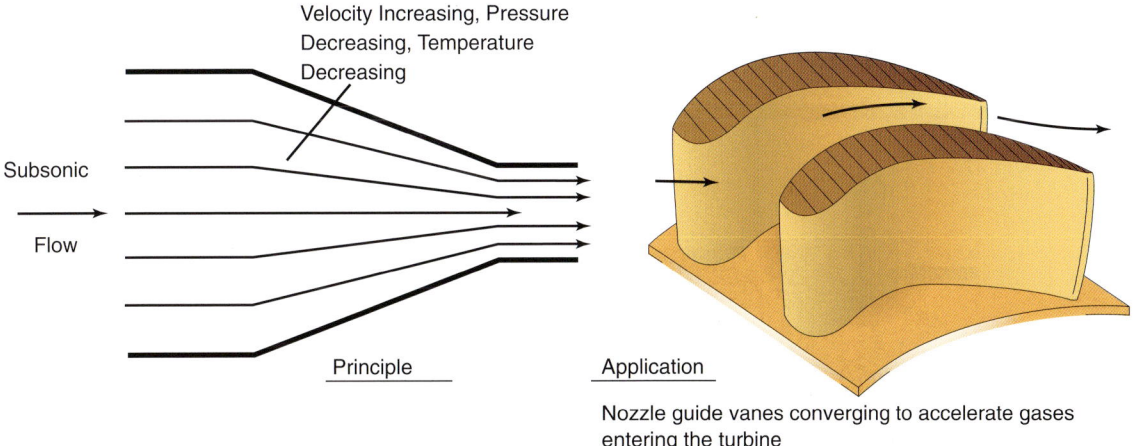

Fig. 11.5

A convergent duct is one that has an area at the inlet greater than the area at the outlet. When air flows through such a duct, velocity increases at the expense of the static pressure and temperature.

DIVERGENT DUCT

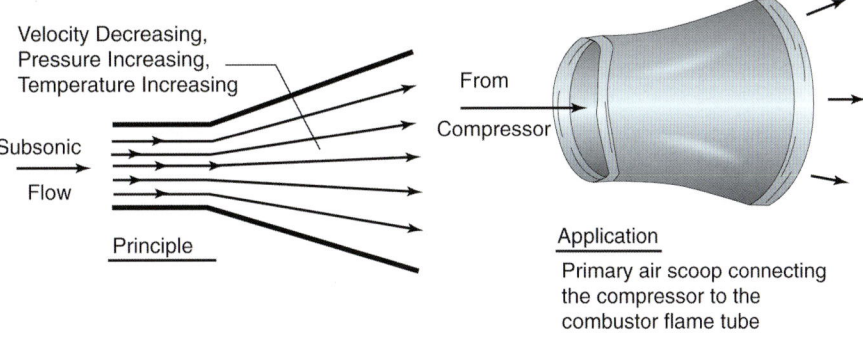

Fig. 11.6

A divergent duct is one that has an inlet area, which is less than the outlet area. This gives a decrease in velocity with an increase in pressure and temperature.

THE WORKING CYCLE OF A GAS TURBINE ENGINE

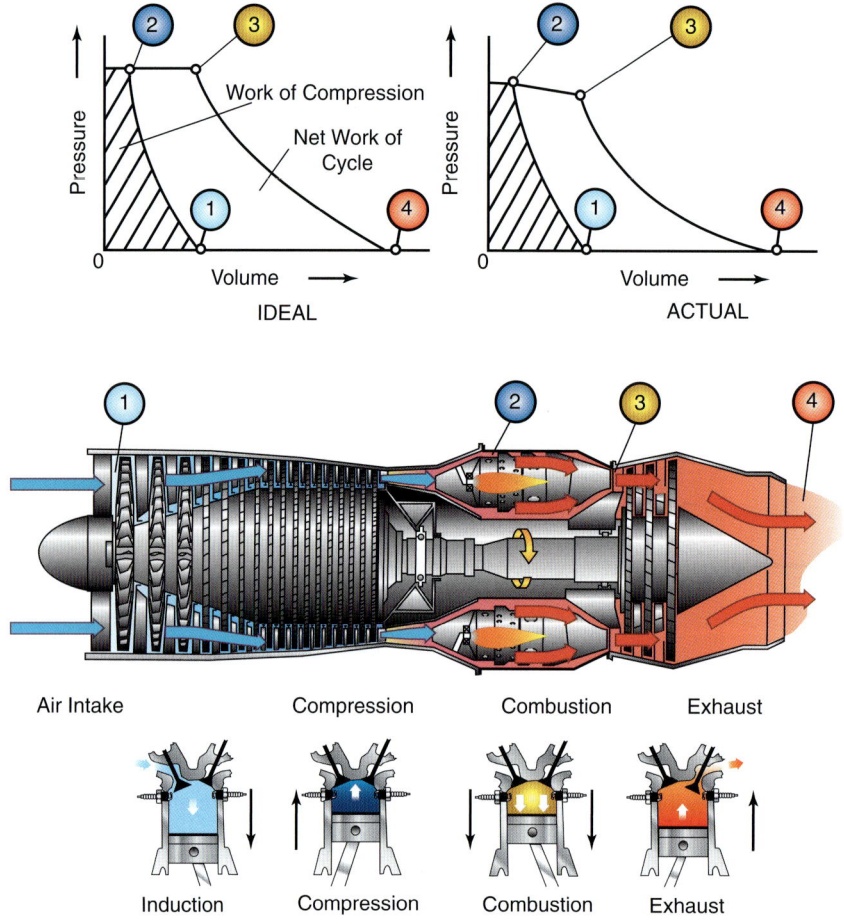

Fig. 11.7

A heat engine converts the heat energy of the fuel into mechanical work. Piston engines and gas turbines are heat engines, both using air as the working fluid.

The "suck, squeeze, bang, blow" (i.e. induction, compression, combustion, exhaust) working cycle of a piston engine is known as the **constant volume cycle** where combustion occurs to give the greatest pressure at the smallest volume. This cycle produces power only on one of the four strokes. This intermittent means of power production compares unfavourably with the working cycle of the gas turbine where combustion is a continuous process resulting in a continuous power output which considerably reduces vibration.

The working cycle of a gas turbine commences with compression where work is done on the air, resulting in an increase in pressure and temperature and a decrease in volume. The cycle continues with the addition of heat energy that increases the temperature and volume while the pressure remains virtually unchanged, hence the term **constant pressure cycle;** its correct name is the **Brayton cycle**.

The gas then expands through the turbine where the turbine extracts energy resulting in a decrease in temperature and pressure while the volume continues to increase. The expansion process completes through the jet pipe nozzle, which provides the jet energy (Thrust), the gas finally reducing to atmospheric pressure. Figure 11.7 shows a Pressure/Volume diagram of a simple gas turbine working cycle.

Note: The term constant pressure cycle can only be applied as long as the engine is operating under a constant set of conditions. Even so, in practice, there is a slight pressure drop in the combustion system due to the turbulence occasioned by the act of combustion.

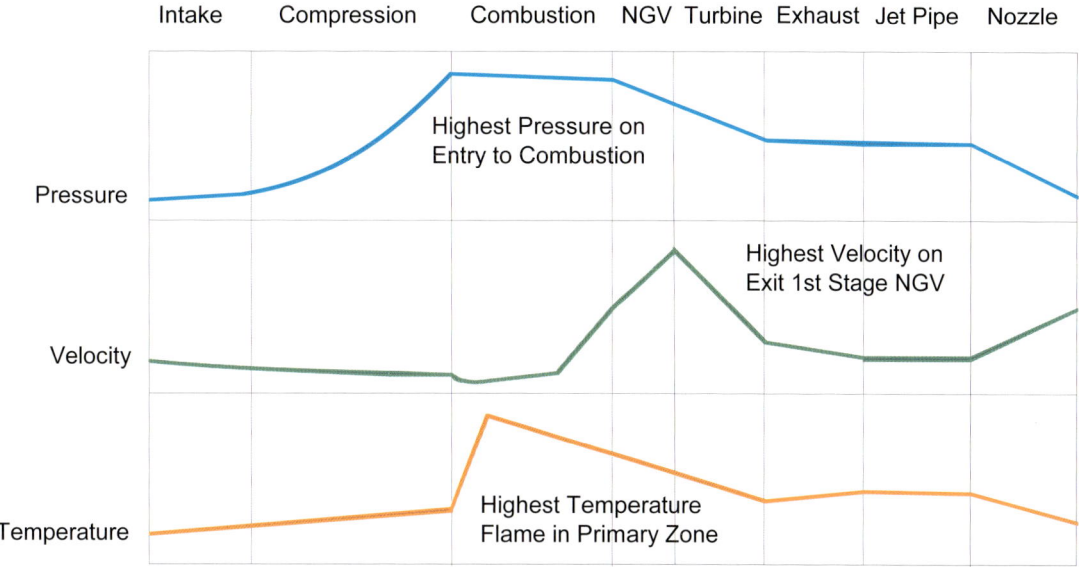

Fig. 11.8

Figure 11.8 illustrates the changes in pressure, velocity, and temperature of the gas flow through the engine.

THRUST

The mass airflow through the engine and the acceleration imparted to it produces thrust. The simple equation is:

$$\text{Thrust (T)} = Ma, \text{ where}$$
$$M = \text{Mass Airflow}$$
$$a = \text{Acceleration}$$

From a simplistic view, in the case of a propeller there is a large mass airflow and a small acceleration, whilst in a gas turbine there is a small mass airflow and a large acceleration. The production of thrust is covered in more detail later under Engine Performance.

POWER

When describing a turboprop, turbo-shaft, or piston engine, the accepted unit for measuring the rate of doing work is **horsepower**. Energy is the capacity for performing work, and power is the rate of doing work. The measure of power is not by the amount of work done, but by units of accomplishment correlated with time. One horsepower is defined as 33 000 foot-pounds of work accomplished in one minute, a foot-pound being the ability to lift a one-pound weight a distance of one foot. Thus, both time and distance are necessary to compute horsepower.

$$P = \frac{F \times D}{T}$$

P = Power
F = Force
D = Distance
T = Time

When a turboprop or a piston engine performs work by driving a shaft that turns a propeller, **torque** and **rpm** can be used to determine the **shaft horsepower (SHP)** that the engine is developing. Torque, in this case, is the twisting or rotary force exerted by the engine to turn the propeller, and rpm is the number of revolutions per minute that the engine crankshaft is making.

EQUIVALENT HORSEPOWER

This is SHP plus the residual jet velocity and can be calculated as net thrust that must be converted to equivalent horsepower (EHP). The formula for calculating equivalent horsepower is as follows:

$$EHP = SHP + \frac{Jet\ Thrust\ lb}{2.6}$$

Where one SHP is equivalent to approximately 2.6 lb of jet thrust.

EFFICIENCIES

In the interest of fuel economy and aeroplane range, the thrust/SHP per unit weight should be at its maximum with the fuel consumption as low as possible.

SPECIFIC FUEL CONSUMPTION (SFC)

This is the amount of fuel burnt per hour of net thrust/SHP, determined by the thermal and propulsive efficiency of the engine.

THERMAL EFFICIENCY

This is the efficiency of conversion of fuel energy to kinetic energy and, like all heat engines, is controlled by the cycle pressure ratio and combustion temperature.

PROPULSIVE EFFICIENCY

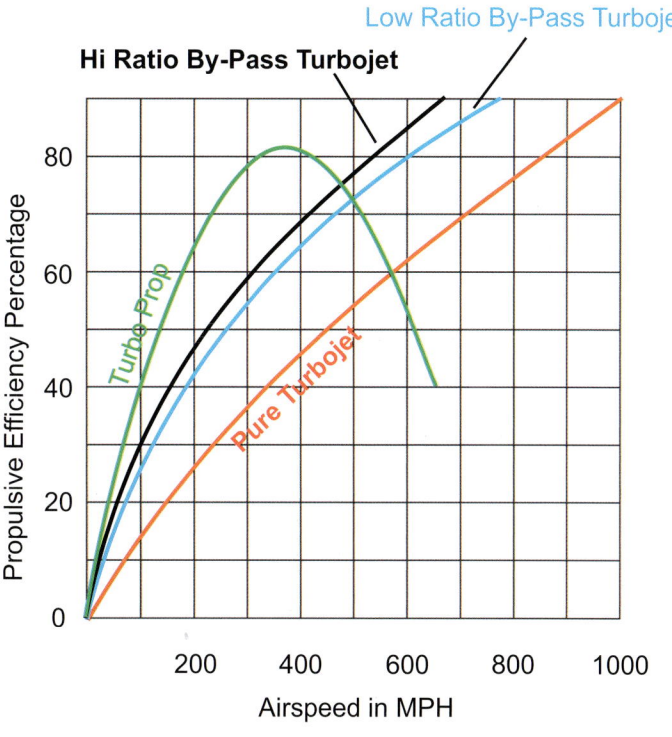

Fig. 11.9

This is the efficiency of conversion of kinetic energy to propulsive work, affected by the amount of kinetic energy wasted by the propelling mechanism.

At aeroplane speeds below approximately 450 mph, the pure-jet is less efficient than a propeller type engine. Since its propulsive efficiency depends largely upon its forward speed, the pure-jet engine is, therefore, more suitable for high forward speeds. The propeller efficiency does decrease quickly over 350 mph. This is due to the disturbance of the airflow caused by the propeller high tip speeds. This has resulted in a departure from pure-jet use where aeroplanes operate at medium speeds by utilising the propeller and gas turbine engine combination.

Introduction of low by-pass ratio turbofan, high by-pass ratio turbofan, and prop-fan has offset the advantages of the propeller/turbine combinations. Figure 11.9 illustrates the differing propulsive efficiencies for the turboprop, high by-pass ratio turbofan, low by-pass ratio turbofan and the pure turbojet.

Chapter 12
Gas Turbine Engines
Types of Construction

TURBOJET

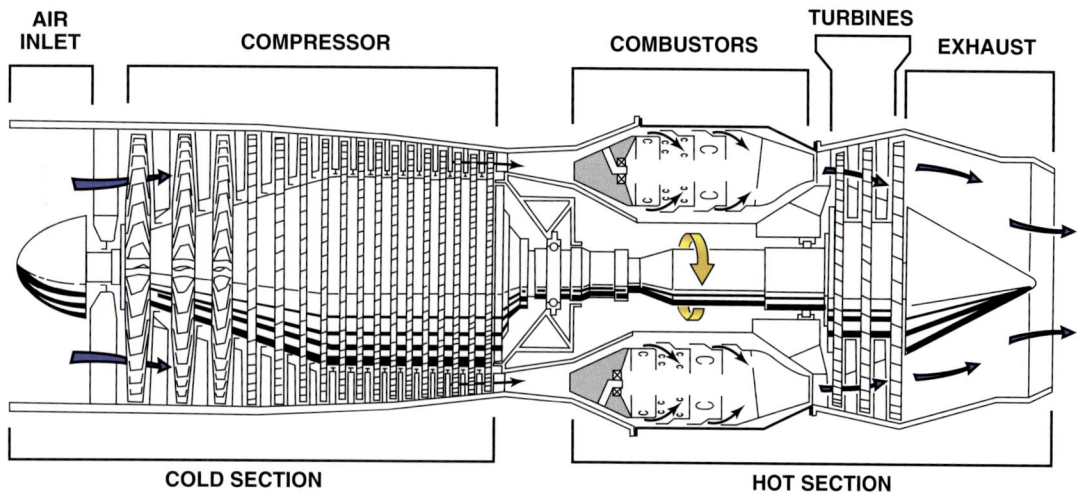

Fig. 12.1 Single Spool Axial Flow Engine

The modern jet engine is cylindrical in shape, as it is essentially a duct into which the necessary parts are fitted. The parts from front to rear include the compressor, the combustion system, the turbine assembly, and the exhaust system. A shaft connects the turbine to the compressor, and fuel burners are positioned in the combustion system. Initial ignition is provided once the airflow is produced by rotation of the compressor. The pressure of the mass ensures that the expanding gas travels in a rearward direction. Initial rotation of the compressor is by means of a starter. Once ignition occurs, the flame is continuous, fuel is supplied, and the ignition device can be switched off. The hot gases crossing the turbine produce torque to drive the compressor. Therefore, the starter can also be switched off. The disadvantages of this type of engine are that they have high fuel consumption and are very noisy, but are efficient at very high speed and altitude.

HIGH BY-PASS TURBOFAN

JT9D-20 TURBOFAN ENGINE

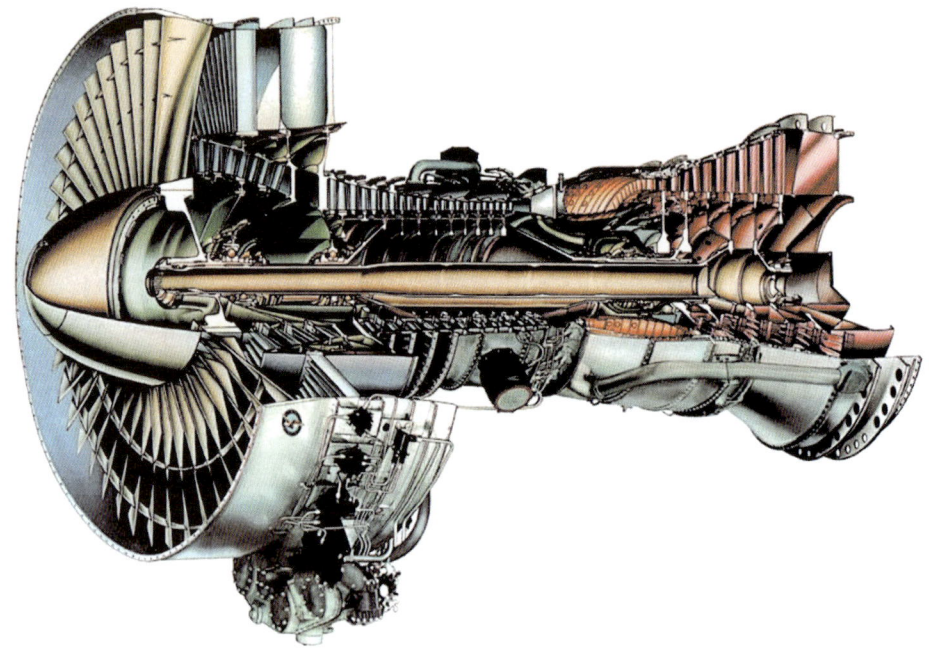

Fig. 12.2

The by-pass engine is fundamentally the same as the turbojet, however, it utilises a duct that surrounds the core engine. The core engine consists of the compressor, combustor, and turbine and in some installations is known as the gas generator section. In addition to the air that flows through the core engine, more air flows through the outer duct.

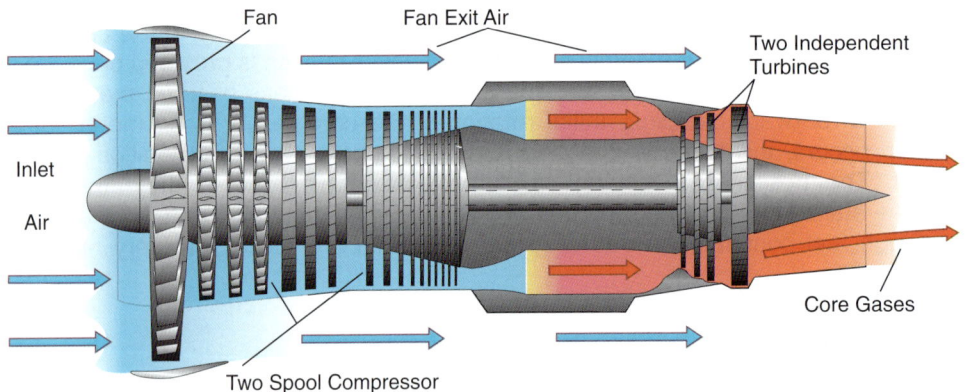

Fig. 12.3

An additional low-pressure (LP) turbine is mounted behind the regular high-pressure (HP) turbine of the core engine. This low-pressure turbine is also driven by the exhaust gases, which give up even more of their energy. The low-pressure turbine is connected to a multi-blade fan at the very front of the engine by a shaft that passes inside the core engine high-pressure (HP) turbine-compressor shaft. This integral shaft is also attached to a low-pressure (LP) compressor. The low-pressure compressor supplies air to the core engine compressor.

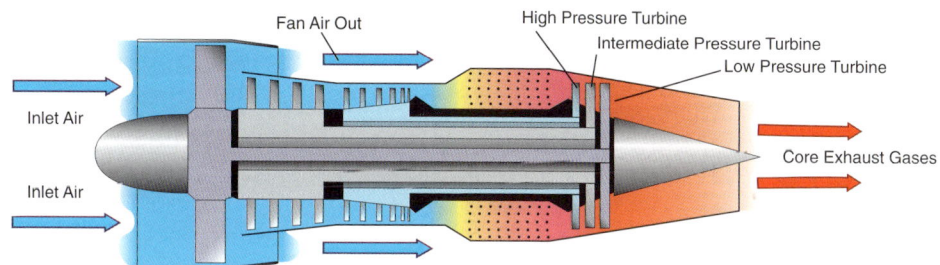

Fig. 12.4

Some high by-pass turbofan engines utilise a three-shaft system. This allows each compressor section to be driven at the optimum rpm. Referring to figure 12.4, the fan links to the LP turbine, the IP compressor links to the IP turbine, and the HP compressor to the HP turbine.

The fan of a turbofan acts like a high-speed propeller and forces air through the duct around the core engine. By using this additional fan, the efficiency of the engine considerably increases at speeds below Mach 1.0, the speed of sound, where straight turbojet engines without fans use more fuel. The mass of air forced backward is greatly increased, and so the thrust of the engine is increased.

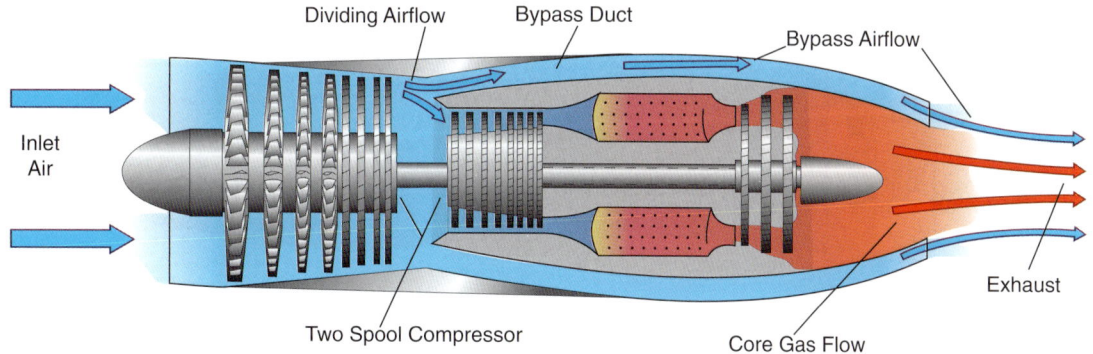

Fig. 12.5

A number that indicates how much more air goes through the duct surrounding the core engine, rather than through the core engine, describes these engines. This number, called the **by-pass ratio**, typically ranges from six to one depending on the design of the engine. Another advantage of the high by-pass engine is reduced engine noise, since the exhaust gases from the core engine are moving at a lower speed relative to the surrounding air (compared to a pure turbine), because more of the energy was used to turn the additional turbine. Less friction exists and less noise is created. Low by-pass engines have a by-pass ratio of up to approximately one to one, see figure 12.5 for a design of a low bypass engine.

TURBOPROP ENGINE

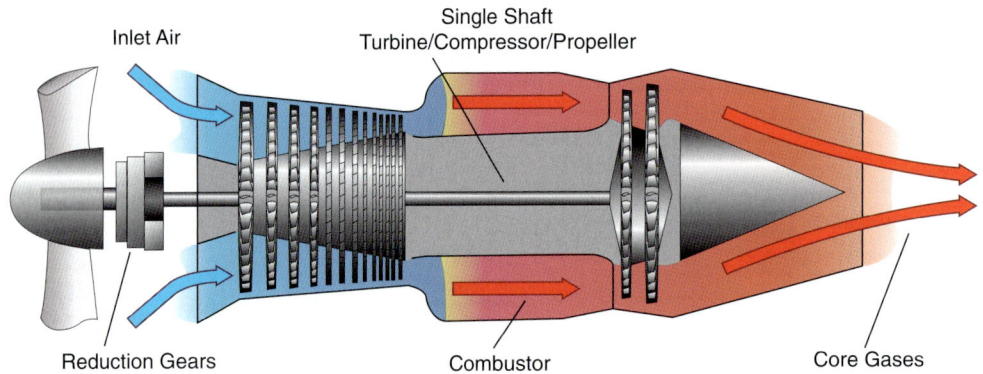

Single Spool

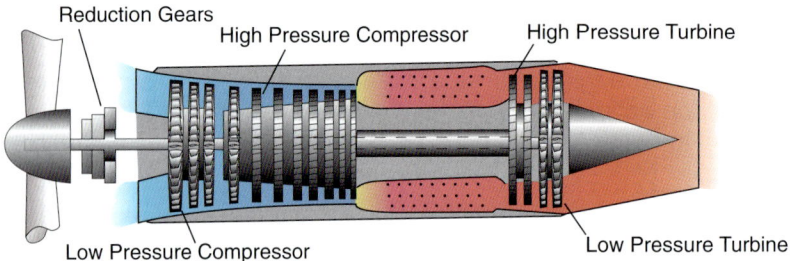

Twin Spool

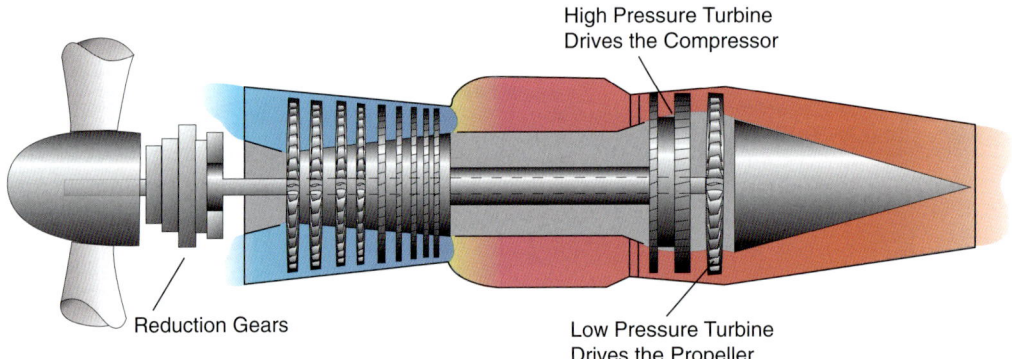

Free Power Turbine

Fig. 12.6

A turboprop engine is nothing more than a gas turbine or turbojet with a reduction gearbox mounted in the front or forward end to drive a standard aeroplane propeller. This type of engine uses almost all of the exhaust gas energy to drive the propeller and provides very little thrust through the ejection of exhaust gases. The exhaust gases represent only about 10% of the total amount of energy available. The turbines extract the other 90% of the energy in driving the compressor and the propeller, either as a direct drive or free turbine.

The basic components of the turboprop engines are identical to the turbojet; that is compressor, combustion section, and turbine. The only difference is the addition of the reduction gearbox to reduce the rotational speed to a value suitable for propeller use.

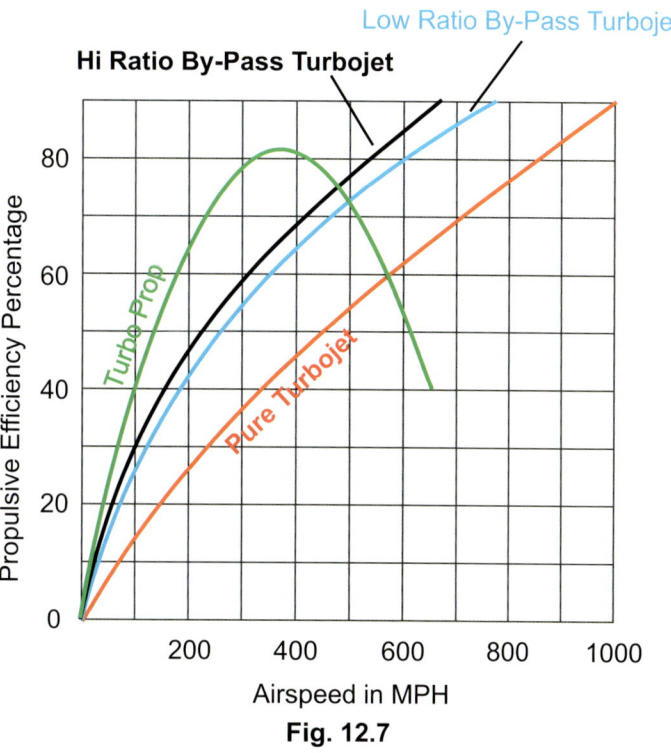

Fig. 12.7

Turboprops are more efficient at low subsonic speeds. However, this advantage decreases with an increase in speed, and since the thrust is derived from the propeller and not the exhaust gas, it has a low noise level. Refer to figure 12.7.

TURBOSHAFT ENGINE

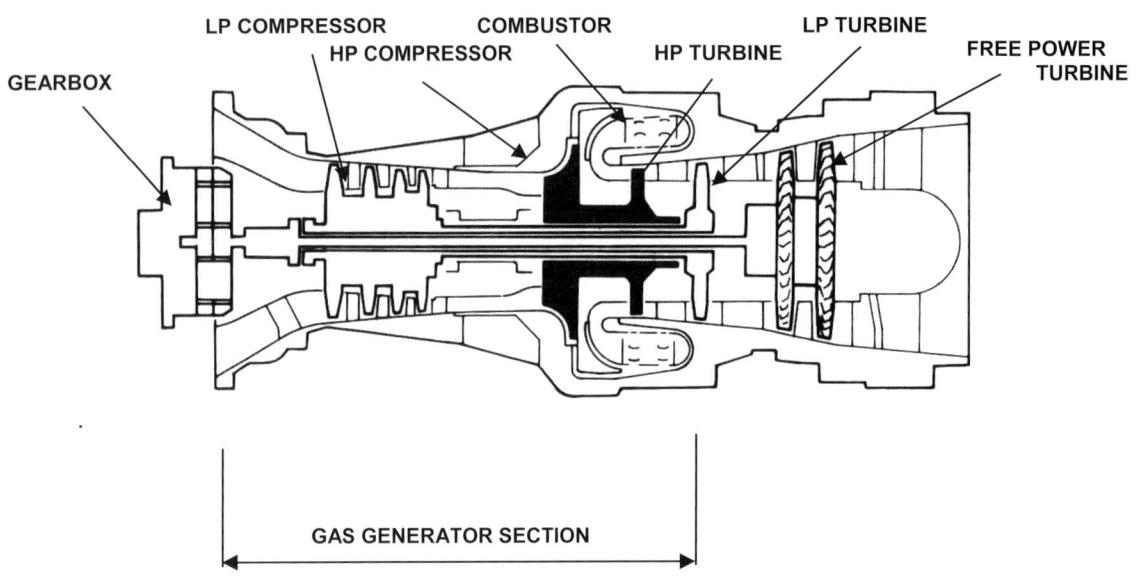

Fig. 12.8

A gas turbine engine that delivers power through a shaft to operate something other than a propeller is referred to as a turboshaft engine. Turboshaft engines are similar to turboprop engines. The power take-off may be coupled directly to the engine turbine, or an independent turbine may drive the shaft called a free power turbine.

The free power turbine is located downstream of the engine's turbine, rotates independently, is not mechanically connected to the main engine, and drives a gearbox rather than a compressor. This principle is used in the majority of turboshaft engines currently produced and is being used extensively in helicopters, hovercrafts, ships, and industrial applications. The main core engine that does not include the free power turbine is called the **gas generator section**.

**Chapter 13
Gas Turbine Engines
Air Inlet**

INTRODUCTION

Fig. 13.1

The function of the air intake is to present a relatively distortion free, high-energy supply of air in the required quantity to the compressor. A uniform and steady airflow is necessary to avoid compressor stall and excessive internal engine temperatures at the turbine. The high energy enables the engine to produce an optimum amount of thrust.

Air inlet design requirements are as follows:

> ➢ The airflow must reach the compressor at a velocity and pressure that enables the compressor to operate satisfactory.
> ➢ It must be able to recover as much of the total pressure of the free airstream as possible and deliver this pressure to the front of the compressor with the minimum loss. This is known as **ram** or **total pressure (ram) recovery**.
> ➢ It must deliver the air uniformly with as little turbulence and pressure variation as possible.
> ➢ The inlet must hold the drag that it creates to a minimum.

SUBSONIC AIR INLET

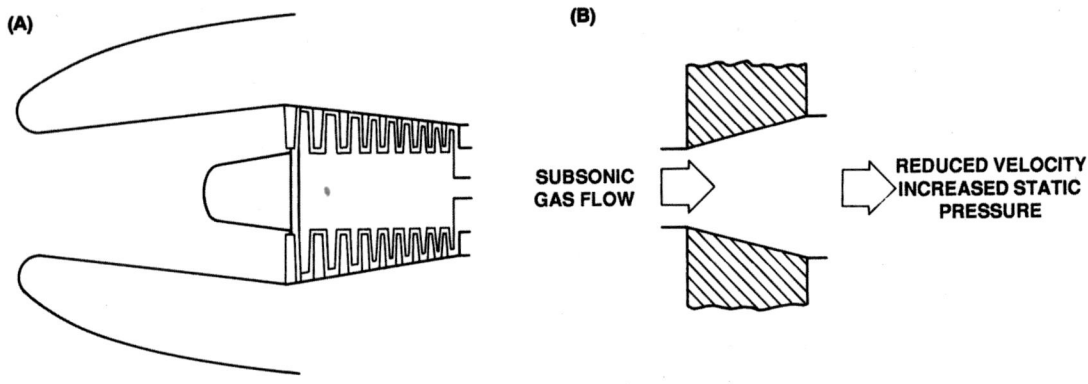

Fig. 13.2

The ideal air inlet for a turbojet engine, fitted to an aeroplane flying at subsonic or low supersonic speeds, is a short, pitot-type circular inlet that is divergent in shape, therefore changing the ram air velocity into higher static pressure. This type of inlet makes the fullest use of the ramming effect on air due to forward speed, suffering the minimum loss of ram pressure with changing aeroplane attitude. However, as sonic speed is approached, the efficiency of this type of inlet starts to fall due to of the formation of shock waves at the intake lip. For a subsonic inlet, airspeed is normally controlled between M 0.4 to M 1.0.

SUPERSONIC AIR INLET

It is necessary at high Mach numbers to have an air inlet with a variable throat area and spill valves in order to accept and control the changing volume of air (e.g. Concorde). This type of intake is described later. Airflow velocities in the higher speed range of the aeroplane are much higher than the engine can efficiently use, so the air velocity must be decreased between the inlet and the engine air inlet.

At flight speeds just above the speed of sound, only slight modifications to ordinary subsonic inlet design are required to produce satisfactory performance. To minimise energy loss and temperature rise at supersonic flight speeds, the inlet is required to slow the air with the weakest possible series or combination of shock waves.

One of the least complicated types of inlet is the simple normal shock type diffuser that employs a single normal shock wave at the inlet with a subsequent internal subsonic compression. Normal shock wave strength at low supersonic Mach numbers is not too high. Therefore, this type of inlet is satisfactory. At higher supersonic Mach numbers, the single normal shock wave is strong and causes a large reduction in the total pressure recovered by the inlet. It is also necessary to consider that the wasted energy of the airstream produces an additional undesirable temperature rise of the inlet airflow.

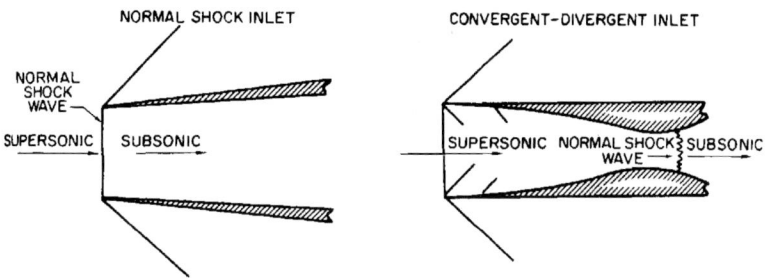

Fig. 13.3

Capturing the supersonic airstream results in ingestion of the shock wave formations, and a gradual contraction reduces the speed to just above sonic. A normal shock wave is produced in the subsequent diverging flow section, which slows the airstream to subsonic. Further expansion continues to slow the air to lower subsonic speeds. This is the convergent-divergent type inlet illustrated in figure 13.3. If the initial contraction is too extreme for the inlet Mach number, the shock wave formation is not ingested and moves out in front of the inlet. The resulting external location of the normal shock wave produces subsonic flow immediately at the inlet, with a resulting greater loss of airstream energy occurring because the airstream is suddenly slowed to subsonic through the strong normal shock.

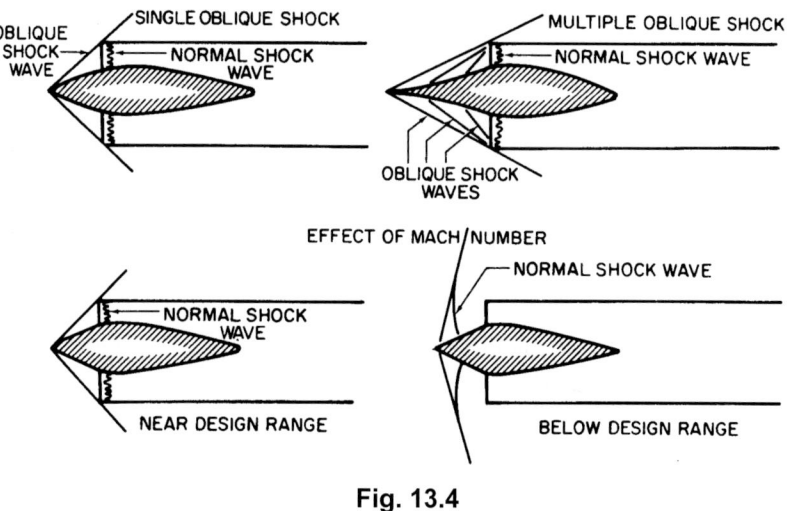

Fig. 13.4

Other diffuser designs, shown in figure 13.4, employ an external oblique shock wave that, before the normal shock occurs, slows the supersonic airstream. Ideally, slowing of the supersonic airstream would gradually occur through a series of very weak oblique shock waves to a speed just above sonic velocity. Then, the subsequent normal shock to subsonic could be quite weak. Such a combination of the weakest possible waves would result in the least waste of energy and the highest pressure recovery. The efficiency of various types of diffusers is shown in figure 13.5, which illustrates this principle.

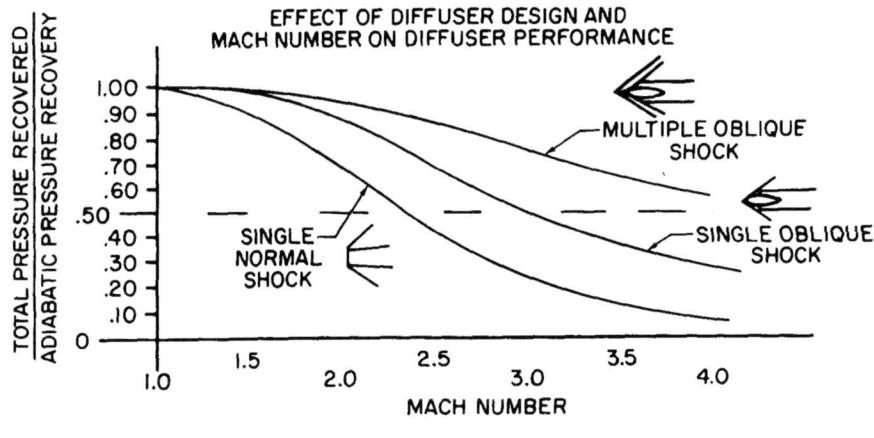

Fig. 13.5

A supersonic inlet complication is that the optimum shape is variable with inlet flow direction and Mach number (i.e. to derive highest efficiency and stability of operation, the geometry of the inlet would differ at each Mach number and angle of attack of flight). A supersonic aeroplane can experience large variations in sideslip angle, angle of attack, and flight Mach number during normal operation, therefore creating certain important design considerations.

> ➢ The inlet should provide the highest efficiency; where the ratio of recovered total pressure to airstream total pressure is a measure of this efficiency.
> ➢ The demands of the powerplant for airflow must be matched by the inlet, and the airflow captured by the inlet should match that necessary for engine operation.
> ➢ At flight conditions other than the design condition, operation of the inlet should not cause a noticeable loss of efficiency or excess drag, and operation of the inlet should be stable, not allowing 'buzz' conditions (i.e. possible during off-design operation oscillation of shock location).

The development of a good, stable inlet design may compromise the performance at the design condition requiring special geometric features for the inlet surfaces or a completely variable geometry inlet design to cater for large variation of inlet flow conditions.

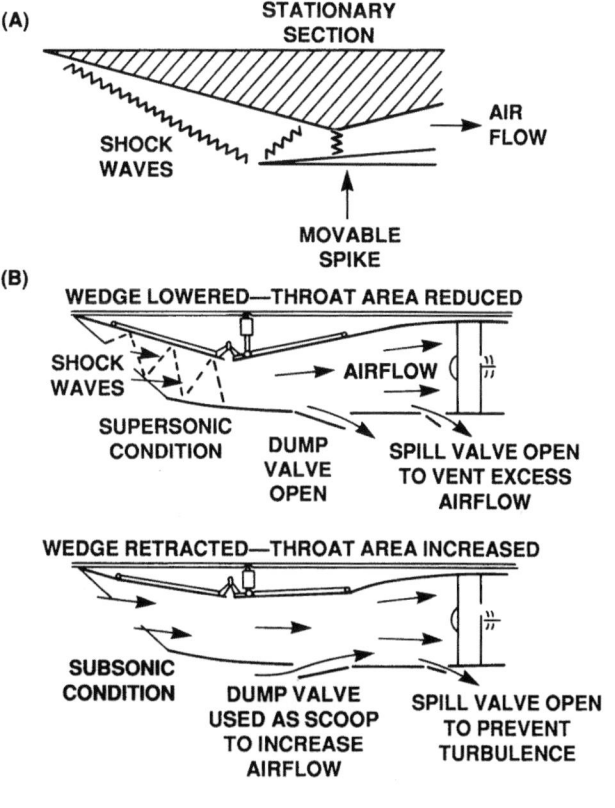

Fig. 13.6

The variable inlet design consists of a variable inlet wedge that reduces or increases throat area, a dump valve situated just after the throat area, and a spill valve located just prior to the compressor.

During subsonic flight, the wedge is retracted to increase the throat area, and the dump valve operates as a scoop to increase the airflow, whilst the spill valve is open, spilling air to prevent turbulence.

In the supersonic flight condition, the wedge is lowered, reducing the throat area, and both the dump and the spill valves are open to vent excess airflow. With changing airspeeds, the variable inlet automatically varies the angle of the inlet.

OPERATIONAL PROBLEMS

Certain operational problems causing disturbed airflow exist during engine operation. Take every care to avoid them. Strong crosswind effects on the inlet, especially on the ground, either sideways, up, or down, may cause eddies in the airflow to the engine and should be corrected by reducing the crosswind effect, such as repositioning the aeroplane. Ice accumulation on the inlet lip or in the inlet can cause airflow disturbances, therefore, switch on anti-icing. Similarly, damage to the inlet can cause airflow disturbances. Heavy in-flight turbulence may trigger a stall under marginal conditions of velocity distribution or engine acceleration.

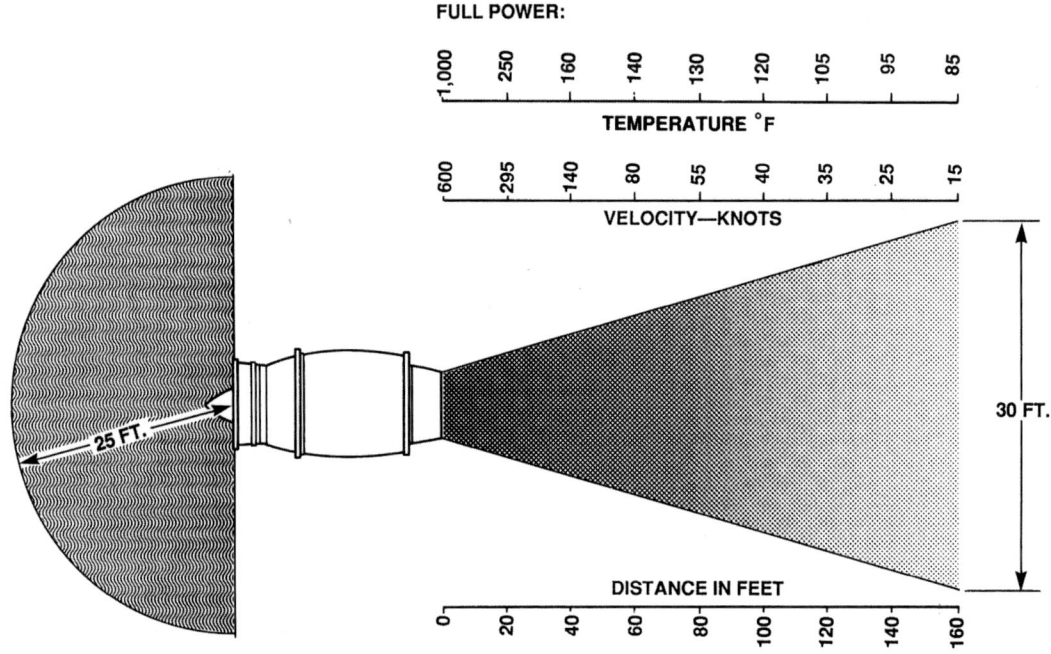

Fig. 13.7

With the development of larger and higher thrust turbojet engines, the danger areas around the engine have become increasingly larger and more hazardous. Items such as dirt, stones, tar strips, nuts, bolts, small tools, rags, hats, clothing, and even the wearer of the clothing can all be ingested into the engine inlet from amazing distances in front, to the sides, and even from partly behind the inlet. The damage caused by material sucked into the engine is known as foreign object damage. Prior to starting the engine, all areas around the engine must be clear of foreign objects, vehicles, equipment, and personnel.

Chapter 14
Gas Turbine Engine Compressors

ACKNOWLEDGEMENT

We would like to thank and acknowledge:

For figures 14.2 and 14.23   Midlands Air Museum

INTRODUCTION

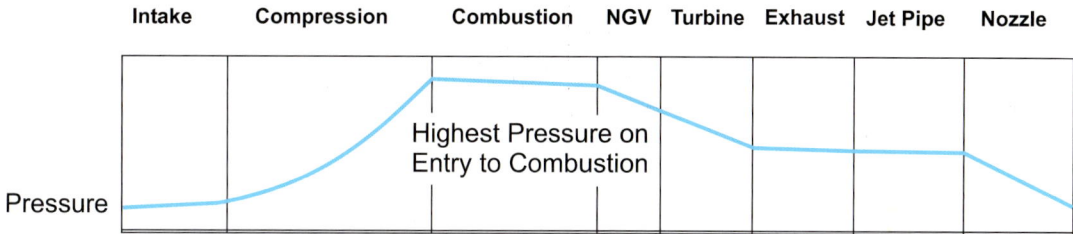

Fig. 14.1

The primary function of the compressor is to increase the pressure of the air mass received from the air intake duct and direct it to the combustion section in the quantities and pressures required to ensure satisfactory operation of the engine.

Fig. 14.2

Two types of compressor are used in gas turbines, centrifugal vane and axial flow. Figure 14.2 shows a double entry centrifugal impeller in the left picture from Frank Whittle's engine, and an axial flow compressor from a Pratt & Whitney PT6.

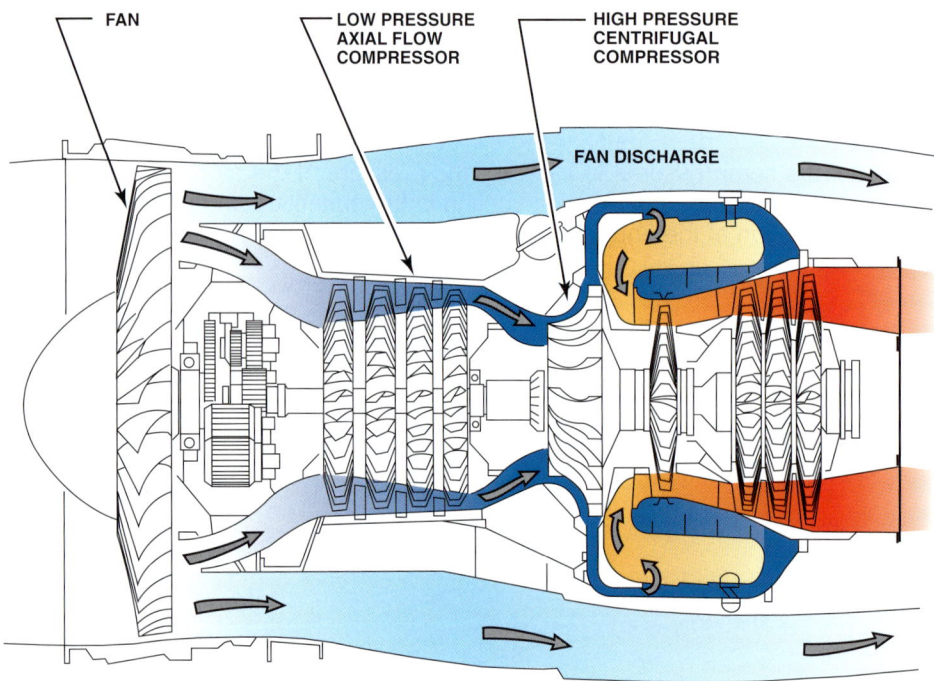

Fig. 14.3

In some engine designs a combination of centrifugal and axial flow compressors are used where the axial flow compressor usually acts as a booster prior to a centrifugal compressor.

With regard to the advantages and disadvantages of the two types, the centrifugal compressor is more robust than the axial flow type. It is also easier to develop and manufacture. The axial flow type, however, consumes far more air than the centrifugal compressor of the same frontal area and can also be designed for high pressure ratios more easily. Since the airflow is an important factor in determining the amount of thrust, this means that the axial compressor engine also provides more thrust for the same frontal area.

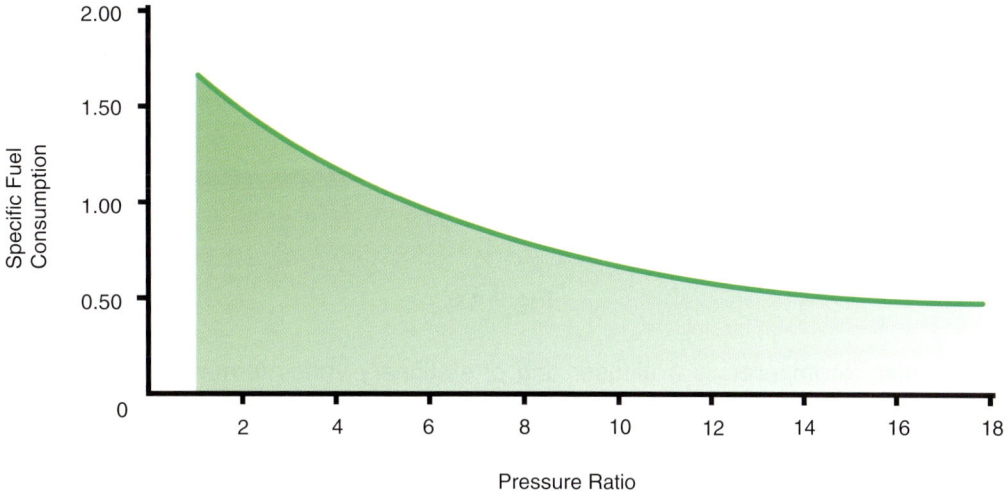

Fig. 14.4

With higher pressure ratios, there is higher engine efficiency and performance due to an improved specific fuel consumption and thrust. A secondary function of the compressor is to provide bleed air for cabin pressure and other aircraft systems. Taking air from the engine affects the pressure ratio (dependent on size of engine and amount drawn off), affecting the engine's overall efficiency.

CENTRIFUGAL COMPRESSOR

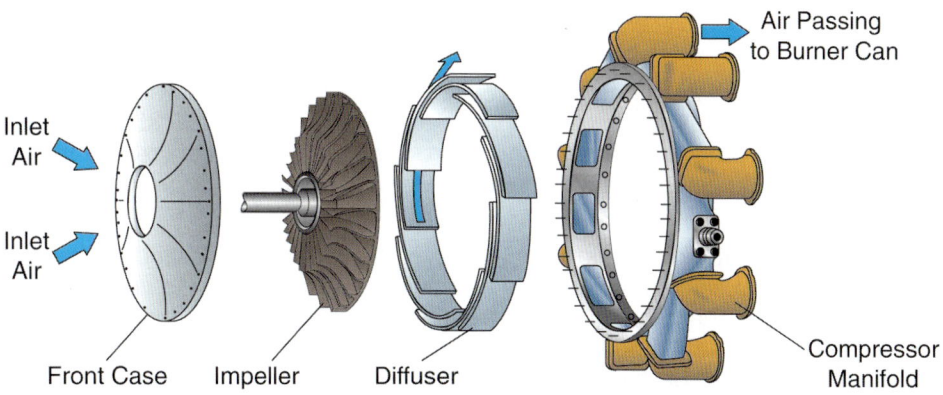

Fig. 14.5

This type is similar to the supercharger used in piston engines, consisting of a disc with a number of radially spaced vanes.

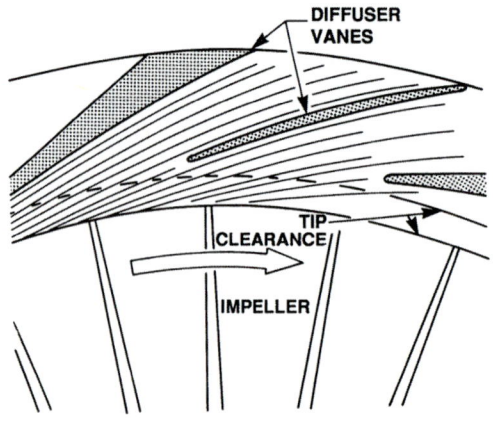

Fig. 14.6

Around the disc, or impeller, is a diffuser ring of stationary vanes formed with divergent cross sections between them. Each impeller and diffuser ring constitutes a stage of compression. Centrifugal compressors may have two stages of compression, but due to design considerations such as weight, no more than two stages are usually incorporated.

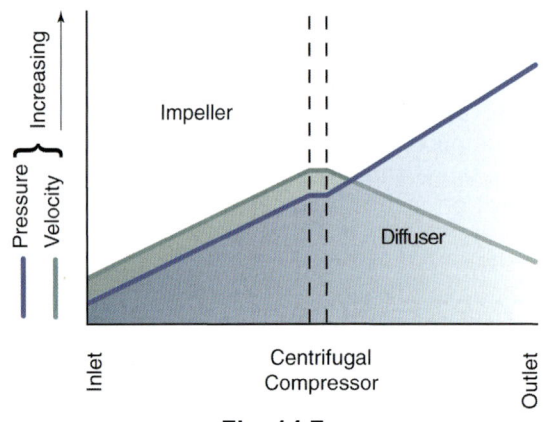

Fig. 14.7

When driven at high speed, the air at the disc centre is forced radially outward along the vanes of the disc. The rotational energy of the disc imparts velocity energy to the air, but because the disc vanes have a divergent passage, some of the energy converts into pressure and temperature. It is normal for the pressure increase through the impeller and diffuser to be approximately equal.

Leaving the impeller tip at high speed, the air enters the **diffuser** ring and passes through its divergent passages. This causes most of the remaining velocity energy to convert into pressure and temperature. It is important that the correct gap between the rotating impeller and the diffuser ring exists, since if the gap is too small aerodynamic buffeting can occur, causing vibration. Additionally, should the gap be too large, a loss of air pressure can occur that decreases the efficiency of the compressor. On leaving the diffuser, the air must turn through 90° in order to enter the combustion section. A divergent duct that has cascade vanes redirects the airflow as smoothly as possible and completes the diffusion.

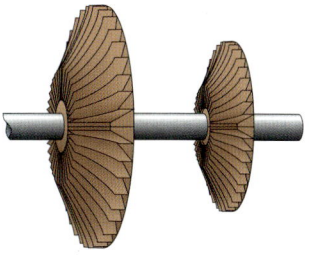

Two Stage

Fig. 14.8

Because of the high rotation speed and drastic changes in the airflow direction, the temperature increase is high; this tends to lower this type of compressor's efficiency. Furthermore, when impeller tip speeds reach sonic values, no further pressure increase is possible. This limits the pressure ratio of this type of compressor to about 4.5:1 for a single stage, and about 6 to 6.5:1 for two stages.

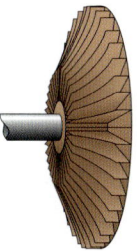

Single Entry

Fig. 14.9

The engine design requirements determine the choice of impeller, but the single entry ducting allows air to be fed into the compressor at the best all round efficiency. Single entry ducting also minimises the chance of surging at altitude, because it makes more efficient use of ram effect than does the double entry ducting. However, to deal with the same mass airflow as a double-sided impeller, a single-sided unit must be of a larger diameter, requiring a lower rotational speed for equivalent tip speeds. A double entry impeller results in pre-heated air entering the engine, since air entering the rear side of the impeller has flowed over the combustion chambers. This can result in unstable operation. Additionally, the turbine must be of a larger diameter to obtain the required turbine blade speed. This increases the engine's overall diameter compared with that of a single-sided impeller type engine, for a given thrust.

In spite of the adoption of the axial flow type compressor, some engines retain the centrifugal type because it:

> ➢ Is simple and comparatively cheap to manufacture
> ➢ Is robust in construction and less vulnerable to damage
> ➢ Gives an almost instant increase in pressure during starting
> ➢ Is less liable to surge

Its main disadvantages are:

> ➢ The high speed of rotation required
> ➢ The large frontal area
> ➢ The limited pressure ratio
> ➢ The high temperature increase

AXIAL FLOW COMPRESSOR

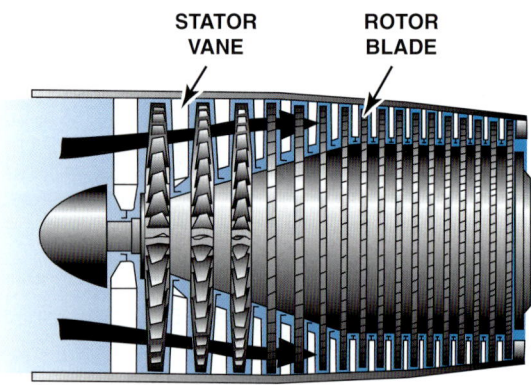

Fig. 14.10

In the axial flow compressor, many stages of moving and stationary blades are necessary. These are positioned alternately, so that each row of rotating (rotor) blades is followed by a row of stationary (stator) blades. A row of rotors and a row of stators form one stage of compression. Because the air leaving each stage is at a higher pressure, it occupies a smaller space. Therefore, each stage of the compressor is smaller than the preceding one, giving the casing a convergent passage, thus maintaining uniform axial velocity.

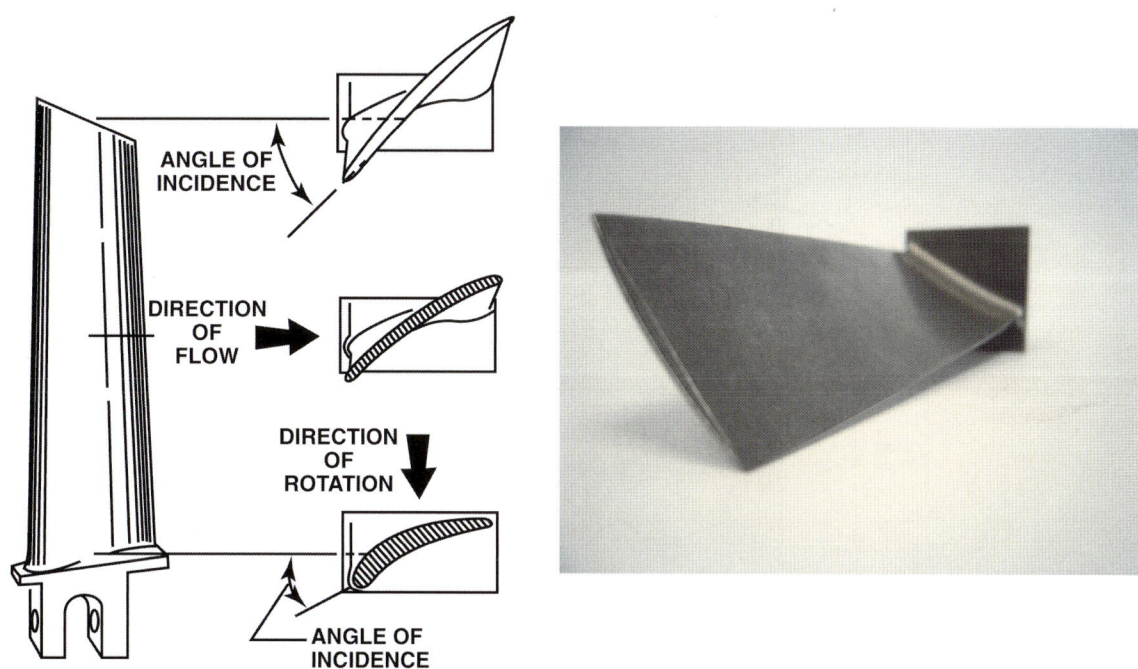

Fig. 14.11

Both the rotors and stators are of aerofoil section. Between each adjacent rotor and stator, the cross-sectional area is **divergent**. During rotation, the rotors act similar to a propeller blade and accelerate the air rearward, converting velocity energy into pressure and temperature. Leaving the rotor, the air passes across the stator. The divergence here causes the remaining velocity energy to convert into pressure and temperature.

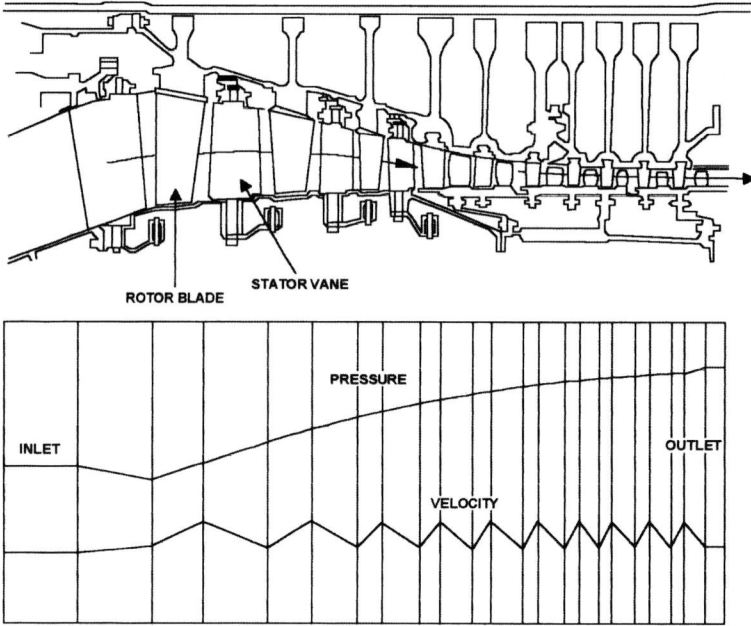

Fig. 14.12

The stators are angled to pass the air into the next stage of the rotor at the correct angle of attack. The final row of stators normally straightens the airflow, removing the swirl of the airflow before it enters the combustion chamber. Each stage increases the pressure by about 1.1 to 1.2:1, and increases the temperature by approximately 25°C. Modern engines can produce compression ratios in excess of 35:1, with compressor outlet temperatures of up to 600°C.

Since the velocity of the rotor blades varies from root to tip it is necessary to twist them to ensure that the airflow maintains a uniform axial velocity along the length of the blade and to give a pressure gradient along their length (see figure 14.11). To relieve stress concentration at the root fixing of the rotor blades, they are loosely mounted. This can cause a clicking sound during windmilling, and is particularly noticeable on a large fan engine.

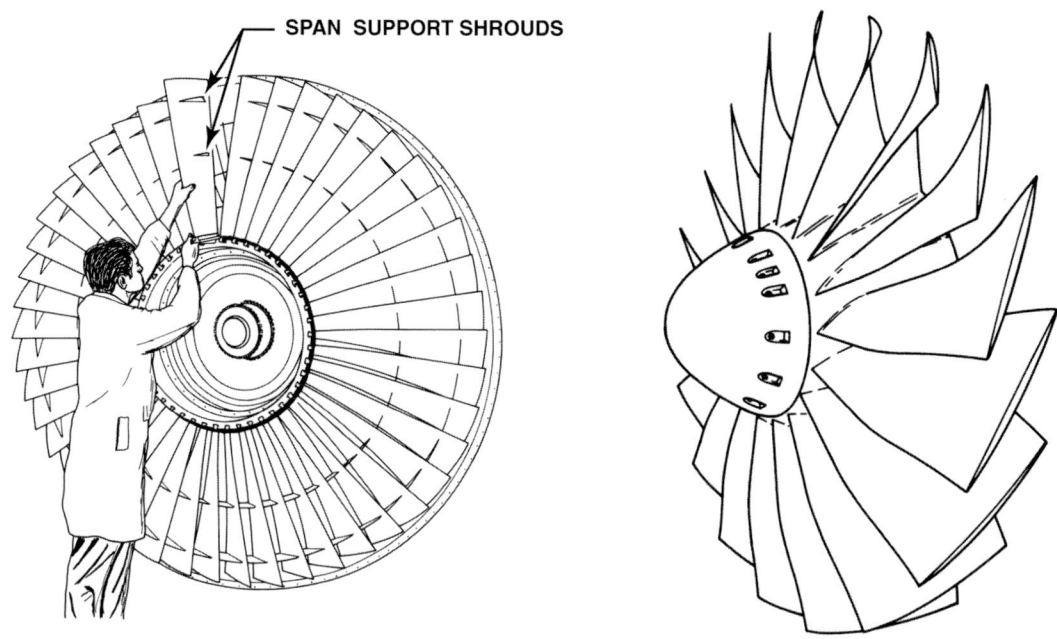

Fig. 14.13

The fan blades of older engines have fitted mid-span supports called snubbers or clappers to overcome resonance vibration that can cause the fan blade to fail. Unfortunately, they are in the most aerodynamically efficient area of the blade. Modern engines have wide chord fan blades that have no snubbers or clappers. They employ a lightweight construction consisting of titanium skins with a honeycomb core. See figure 14.13.

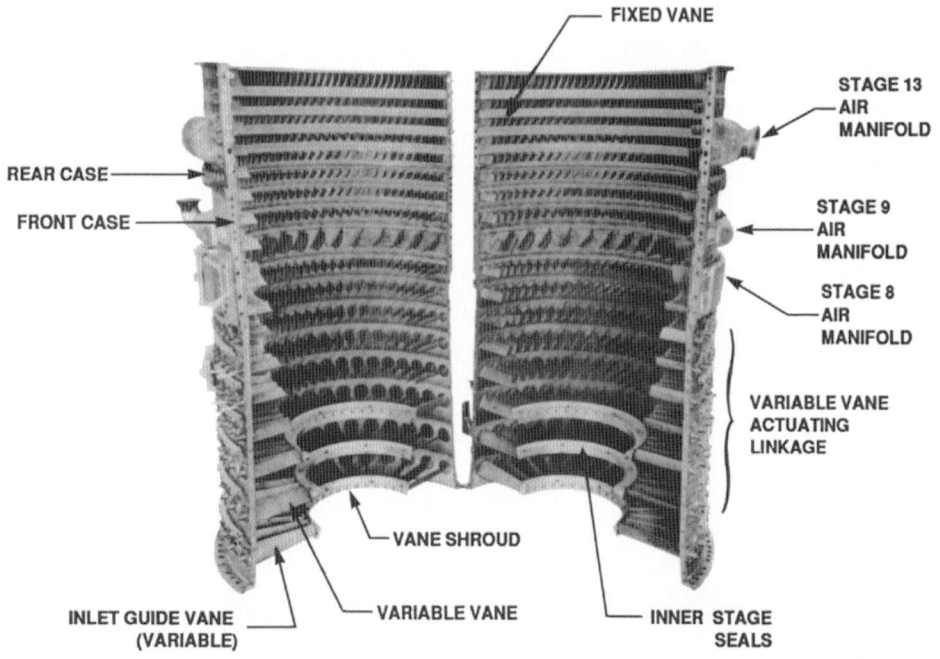

Fig. 14.14

The stator blades are usually mounted in segments and secured either to the engine casing directly or in a retaining ring. The stators toward the front of the compressor may be shrouded at their inboard end to minimise vibration.

Inlet guide vanes are installed before axial flow compressors and may be fixed or variable. Their purpose is to direct the airflow to the rotating assembly at an acceptable angle. Note that inlet guide vanes are not installed before the fan of a turbofan engine.

TYPES OF AXIAL FLOW COMPRESSOR

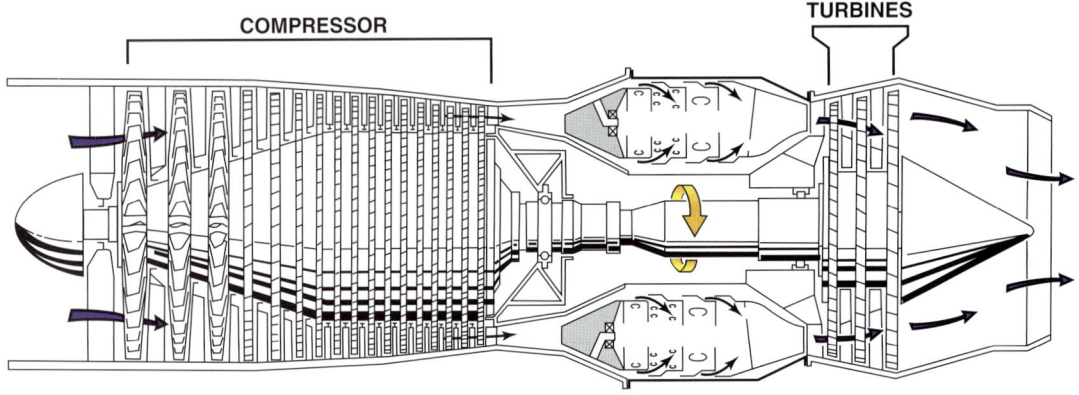

Single Spool

Fig. 14.15

The simplest type of axial flow compressor is the single-spool compressor. It consists of one rotor assembly and stators, with as many stages as necessary to achieve the desired pressure ratio, and all the airflow from the inlet passes through the compressor. Because of the large number of stages required to produce a high compression ratio, as the number of stages increases, it becomes more difficult to ensure that each stage operates efficiently over the engine speed range. To produce higher pressure ratios, it becomes necessary to incorporate a large number of stages.

Some relief from the surging troubles present on a multi-stage, high performance compressor is possible via use of various anti-surge devices. A description of the operation of these devices appears later.

TWIN-SPOOL AXIAL FLOW COMPRESSOR

To achieve greater flexibility of operation, make the engine more efficient, and achieve a higher pressure ratio, multi-spool compressors have been developed consisting of two or more rotor assemblies, each driven by their own turbine at optimum speed.

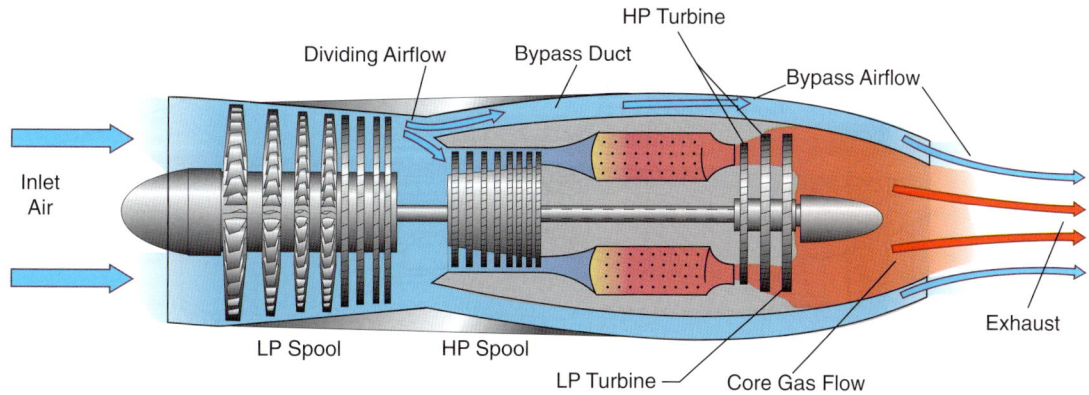

Fig. 14.16

The compressor is made up of two sections. Each section or spool is a completely independent unit, driven by a separate turbine assembly through co-axial shafts. The rear low-pressure (LP) turbine drives the forward, low-pressure (LP) compressor, while the front high-pressure (HP) turbine drives the rear, high-pressure (HP) compressor. This allows the two assemblies to run at different speeds. Figure 14.16 shows a low by-pass twin-spool turbojet.

When operating at the optimum conditions of mass flow and pressure ratio, the performances of the low- and high-pressure assemblies match. As rpm or inlet density changes, the two independent compressor sections adjust their speeds to suit the prevailing airflow conditions. For example, when the engine throttles back, or is running at low rpm, the HP turbine is doing more work than the LP turbine. Consequently, the rpm of the LP compressor decreases relative to that of the HP section. This reduces the velocity of air through the LP section, which allows the angle of attack to the first stage rotor blades to remain virtually unchanged. This thereby decreases the tendency to stall. At the same time, the continuing relatively high rpm of the HP section still allows the maintenance of an effective pressure rise. This arrangement permits the achievement of a high pressure ratio without the penalty of starting or acceleration problems.

Additionally, as altitude increases, the reduction in air inlet pressure causes the rpm of the LP compressor to increase, resulting in an increased airflow to the HP section. Due to this increase, the overall decrease in mass flow is not as great as that through a single spool compressor. Therefore, the decrease in thrust with altitude is also less than that of a single-spool compressor.

Only a percentage of the air from the LP compressor passes into the HP compressor; the remainder of the air, the by-pass flow, ducts round the HP compressor. Both flows mix in the exhaust system before passing to the propelling nozzle.

TURBOFAN (HIGH BY-PASS) TWIN/TRIPLE SPOOL

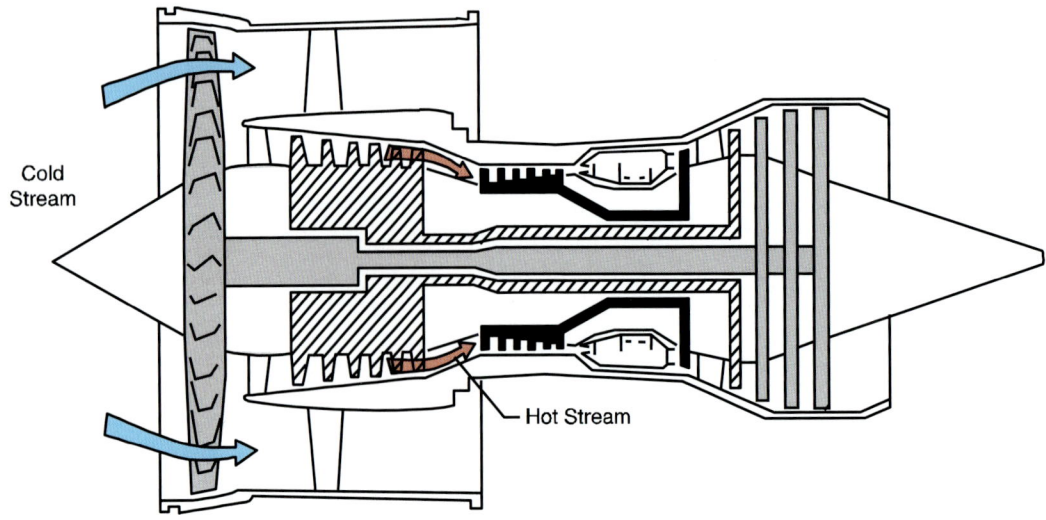

Fig. 14.17

A fan may be fitted at the front of a twin- or triple-spool compressor and, on the twin-spool type of engine, the fan turns at the same speed as the compressor to which it is attached. A large proportion of the air from the outer part of the fan, known as the cold stream, bypasses the core engine and ducts to the atmosphere through the cold stream nozzle, producing the majority of the thrust. The smaller airflow, from the inner part of the fan and known as the hot stream, passes through the other compressors and is compressed further before passing to the combustion system and the turbines.

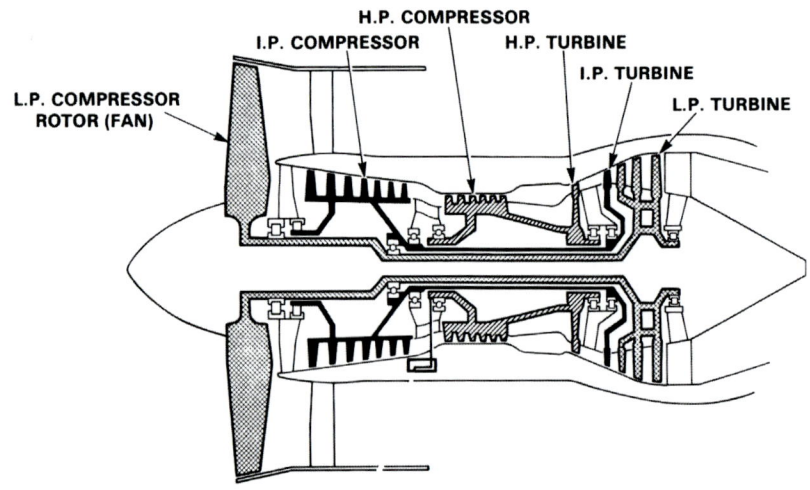

Fig. 14.18

Figures 14.17 and 14.18 illustrate engines of the triple-spool type, where the fan is, in fact, the LP compressor and is driven by its own turbine separately from the intermediate pressure (IP) compressor and the HP compressor. The LP compressor has large rotor (fan) and stator blades, and is designed to handle a far larger airflow than the other two compressors, each of which has several stages of rotor blades.

COMPRESSOR RPM INDICATION

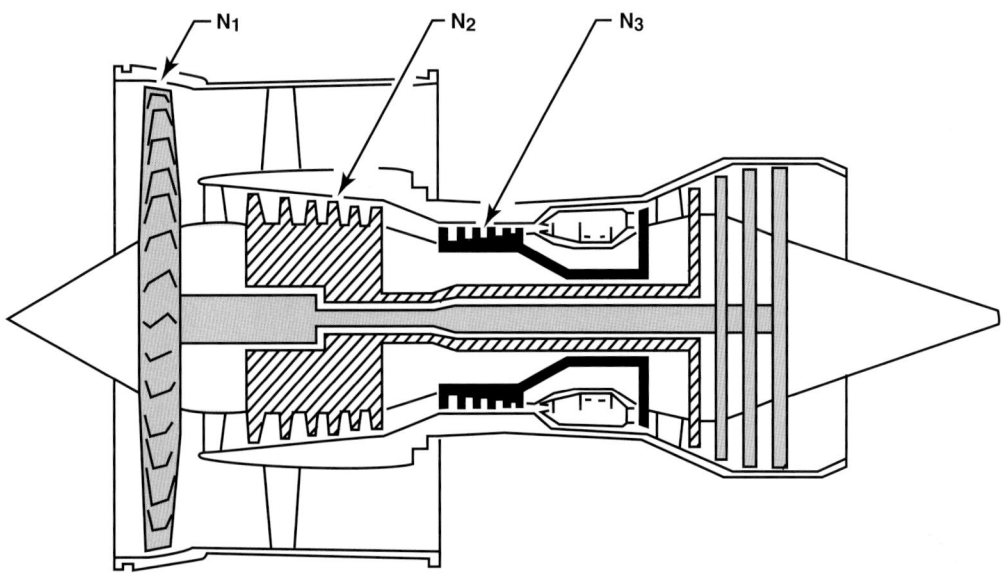

Fig. 14.19

Compressor rpm is indicated as **N**, with each compressor of a multi-spool engine identified by a number. For a twin-spool engine, the LP compressor is identified as N_1 and the HP compressor as N_2. In the case of a triple-spool engine, the LP compressor is N_1, the IP compressor N_2 and the HP compressor N_3.

DIFFUSER

After the air leaves the compressor, its velocity decreases to a value suitable for the combustion process. This is necessary since excessive airspeed within the combustion section makes it difficult to establish and maintain a flame.

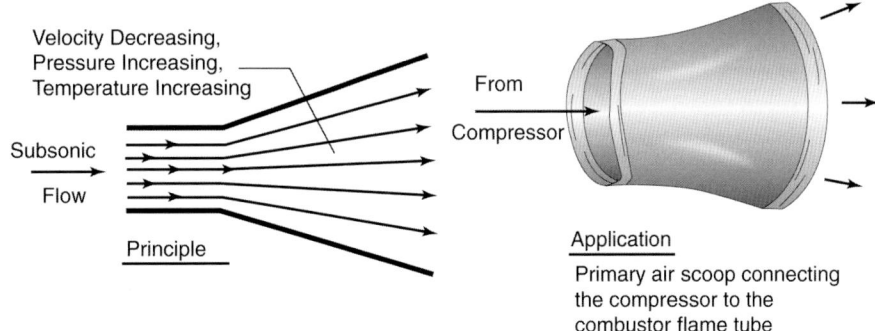

Fig. 14.20

To achieve a reduction in air speed, the direction of air is through a divergent section called a diffuser. As the air leaves the compressor and enters the diffuser section, it has a higher static pressure and total pressure than it had at the entrance of the compressor. As air leaves the diffuser and enters the combustion chamber zone, its velocity relative to the engine is quite low. Its static pressure is almost equal to its total pressure, roughly the same at the outlet of the compressor, with the negligible difference due to air friction.

COMPRESSOR STALL AND SURGE

The airflow characteristics of each stage of a multi-stage compressor are different from those of its neighbour, and each stage must be carefully matched to the next stage to produce an efficient compressor. Failure to do so may result in stalling of groups of blades, individual blades, or stages and may lead to surge where the whole compressor stalls.

The compressor is most efficient when the airflow meets the rotor and stator blades at the correct angle of attack. The angles of attack of the rotor blades are set by the design mass flow, pressure ratio, and rpm, so the most efficient angle of attack through each successive stage is achieved only when the compressor is running under its optimum set of conditions. However, as the engine must operate over a wide range of conditions, the efficient matching of compressor stages becomes critical.

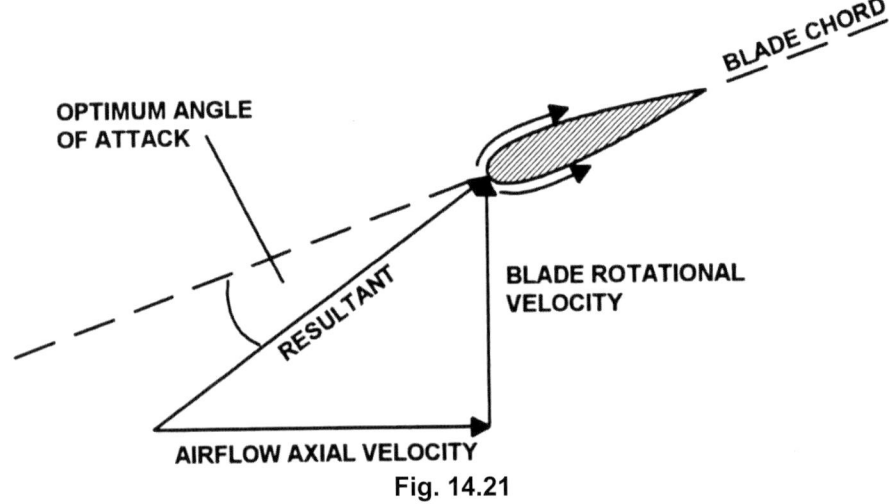

Fig. 14.21

Below the design conditions (i.e. when the relationship between pressure, velocity, and rpm is disturbed), the angle of attack of the early compressor stages becomes too great, and the airflow across the blade breaks away and creates eddies until the blade stalls in the same way as any other aerofoil. The stall may spread downstream to the subsequent stages until the entire airflow pattern breaks down.

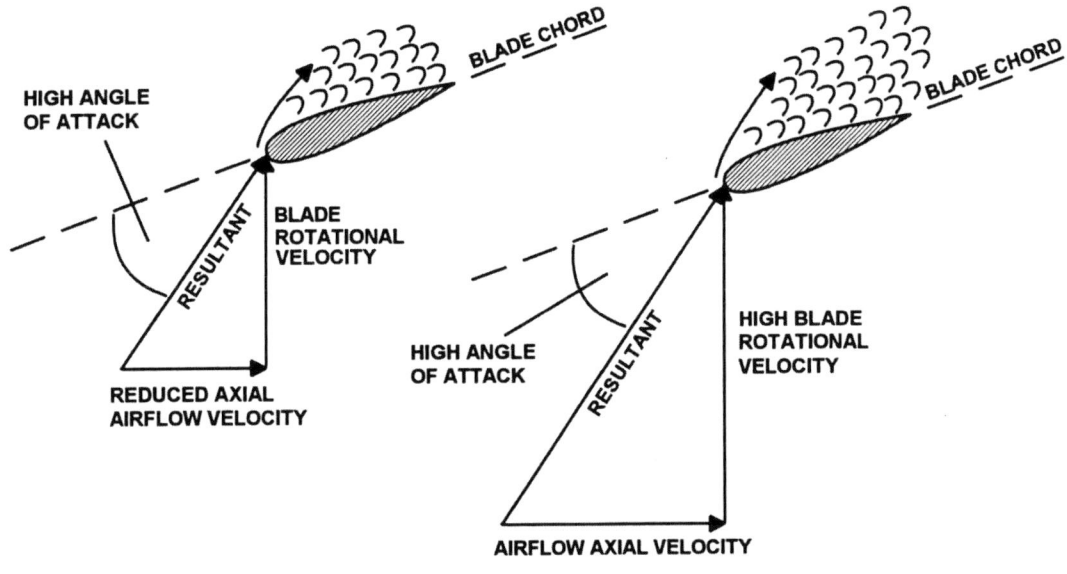

Fig. 14.22

At low engine speeds, another factor can affect the later stages of the compressor. Due to the reduced pressure ratio, the air attempts to occupy a greater volume. However, as the volume of the annular air space controls this, the result is a choking of the later compressor stages. When choking occurs, the velocity of the air flowing through the compressor decreases until the first stage stalls, followed by subsequent stages. This permits a reversal of airflow. Choking is now relieved, restoring the normal airflow pattern until a choked condition re-occurs. This instability of airflow is **surge**. If surging continues, it may result in severe compressor damage necessitating immediate engine shut down. In severe cases, a complete reversal of flow with flames and ejection of engine parts from the inlet can occur.

OVER-FUELLING SURGE

Before considering surge through over-fuelling, it is necessary to analyse the effect of normal throttle opening.

If the engine is running in the lower rpm range and the throttle opens at the normal rate, the increased fuel flow increases the gas temperature and pressure. This results in increased power at the turbine and acceleration of the rotating assembly takes place.

If the throttle opens rapidly, resulting in an excessive rate of over-fuelling, the lag in acceleration response of the rotating assembly (due to its inertia) combined with an excessive increase in the gas velocity causes the turbine to choke. The choking of the turbine decreases the air velocity through the compressor to such an extent that the early compressor stages stall, thereby inducing a surge.

To overcome this surge condition, some turbojet fuel systems have an over-fuelling control, which regulates the over-fuelling rate to match the lag in acceleration of the compressor. It is generally called the acceleration control unit or ACU.

SURGE CONTROL

Using a system of airflow control ensures efficient operation of a gas turbine engine over a large rpm range, and retains the safety margin between normal operating conditions and those conditions causing compressor stall and engine surge. This system usually consists of a row of inlet guide vanes arranged so that their angle adjusts automatically via a control that is sensitive to engine speed, to prevent or minimise compressor stall in the first inlet stage. To improve the smooth airflow through the compressor, valves are fitted to bleed away air from selected intermediate stages of the compressor to the atmosphere. In addition to bleed valves and variable angle intake guide vanes, some engine designs include variable angle stator blades. Variable angle stator blades are automatic in action and fitted to the compressor stages most likely to stall.

BLEED VALVES

Bleed Valve Assembly

Fig. 14.23

Bleed valves control the amount of air bled from the compressor and are open at idle rpm and low pressure ratios, and automatically close at higher rpm (about 85%) when the airflow conditions are more stable. This is because the conditions most likely to bring about compressor stall and engine surge are those encountered when the engine is idling in-flight or during maximum acceleration.

Ice formation in the air intake and certain aeroplane manoeuvres may also induce compressor stalls. The air bleed valves have done the most to reduce the risk of compressor stalls, as during the critical periods these valves are open to decrease the airflow over the intermediate stages. This prevents choking of the rear stages with subsequent stall at the front and engine surge. These valves can operate hydro-mechanically, electrically, or, as shown in figure 14.24, pneumatically.

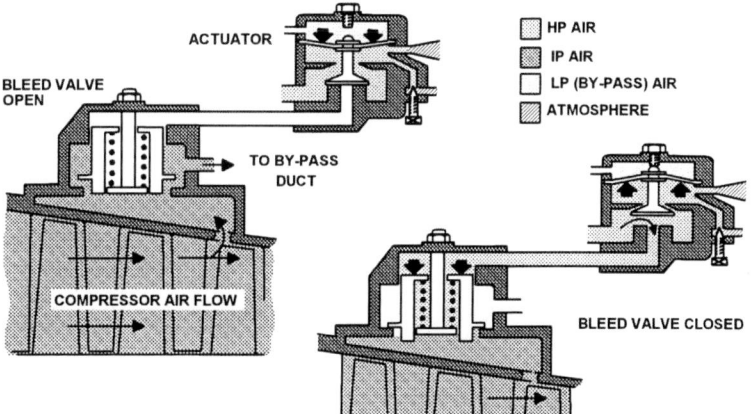

Fig. 14.24

VARIABLE INLET GUIDE VANES

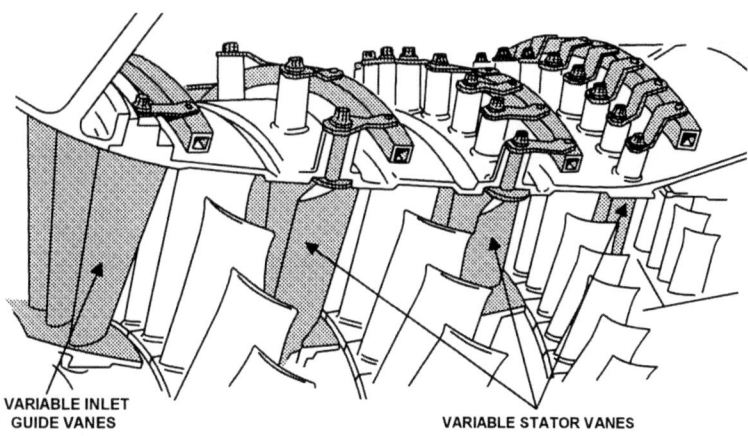

Fig. 14.25

Variable inlet guide vanes, located prior to the first stage rotor, automatically adjust to redirect the airflow into the compressor as compressor speed falls off. Thus, the incoming airflow continues to meet the first stage rotors at the correct angle of attack. The variable angle inlet guide vanes give a swirl to the air entering the front of the compressor, and the amount of swirl adjusts to suit engine running conditions. An actuator and suitable linkage alter the angle of the inlet guide vanes mechanically. The actuator control is sensitive to engine rpm and inlet temperature and the actuators operate hydraulically by fuel pressure from the engine fuel system. Some engine designs also incorporate a number of variable stators that work in conjunction with variable inlet guide vanes. Figure 14.25 illustrates this arrangement.

SURGE ENVELOPE

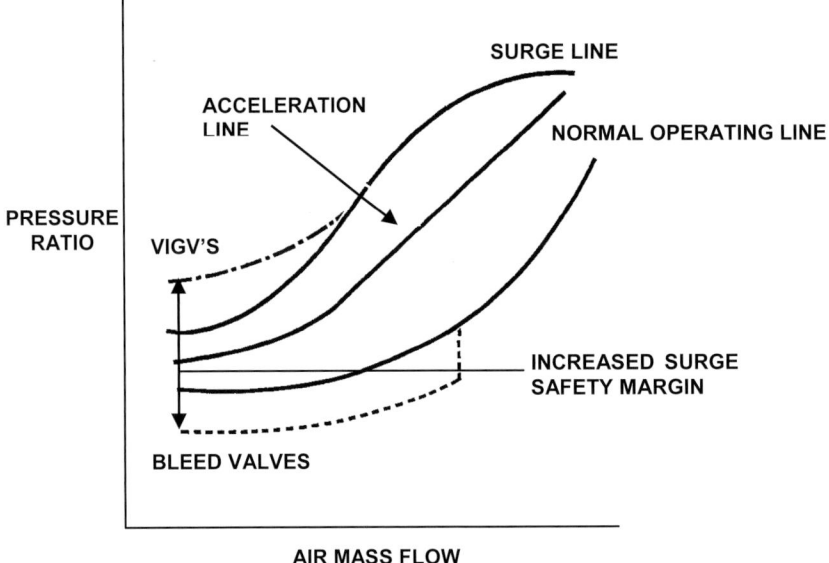

Figure 14.26

Figure 14.26 shows a typical surge envelope that is a result of plotting pressure ratios against air mass flows over a range of rpm. It identifies the surge line, acceleration line, normal operating line (working line), and the effects of the variable inlet guide vanes and bleed valves. The variable inlet guide vanes effectively raise the surge line and the bleed valves lower the normal operating line at low air mass flows (i.e. low engine rpm, resulting in an increase in the surge safety margin).

CAUSES AND INDICATIONS OF STALL

The following conditions are causes for stall and surge:

  ➢ Rapid increase in fuel flow during increase of rpm
  ➢ Low engine rpm (e.g. idle)
  ➢ Strong crosswind on the ground
  ➢ Engine inlet icing
  ➢ Contaminated or damaged compressor blades
  ➢ Damaged engine air inlet

Visual and audible indications of surging are as follows:

  ➢ Thrust loss
  ➢ Erratic rpm
  ➢ Vibration
  ➢ Inability of the engine to accelerate
  ➢ Fluctuating and a rapid rise in exhaust gas temperature
  ➢ Banging and/or rumbling from the compressor
  ➢ In extreme cases, burning gas coming out of the inlet and exhaust

Care must be taken during operation to control the conditions that can result in engine stall or surge. The pilot must carry out the correct procedures to counter any adverse operating conditions. The most common action is to decrease fuel flow by retarding the throttle. Should the problem persist, it may be necessary to shut down the engine.

Chapter 15
Gas Turbine Engine Combustion Systems

INTRODUCTION

The combustion system is designed to burn the fuel as efficiently as possible over the whole range of engine operating conditions. It must do this without any increase in pressure, with all the energy released by the fuel converted into heat and velocity energy.

In addition to the above, the functions of the combustion chamber are as follows:

> To ensure a stable fuel burn with maximum thermal efficiency over a wide range of air pressure, temperature, and mass flow
> To ignite and maintain combustion of air/fuel ratios ranging from 45:1 to 130:1
> To give an even outlet temperature distribution
> To have low overall drop in total pressure
> Be of small size and weight, and have a long mechanical life

COMBUSTION PROCESS

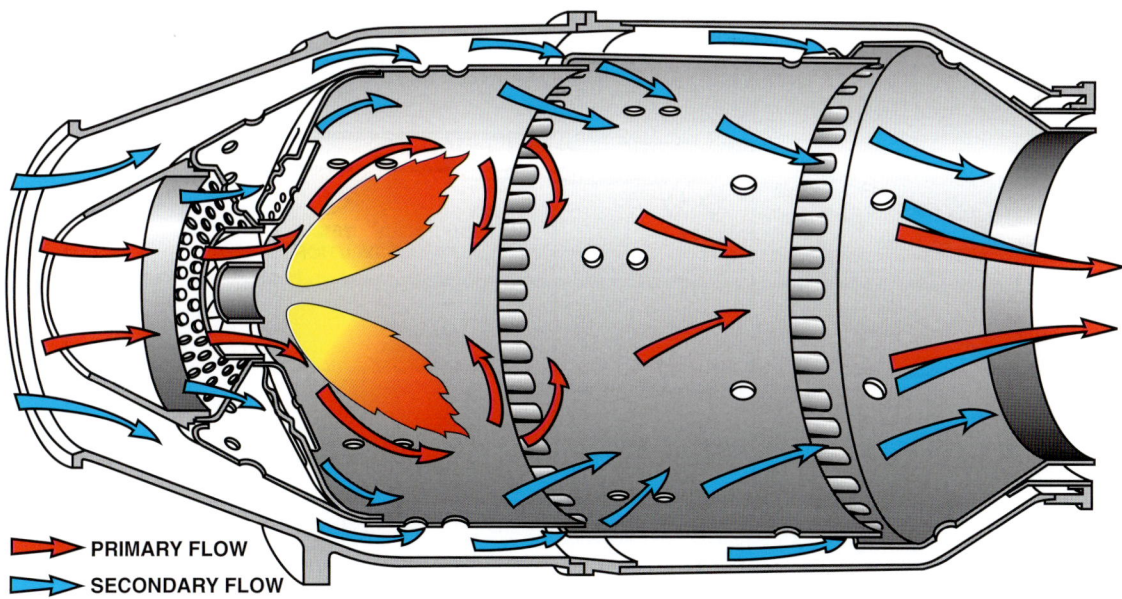

PRIMARY FLOW
SECONDARY FLOW

Fig. 15.1

Air from the compressor enters the combustion system at a velocity of up to 500 ft/second. However, this airspeed is too high to effect satisfactory combustion, so the air must be decelerated, which increases the static pressure.

Efficient combustion only takes place within a narrow band or range of air/fuel ratios of approximately 15:1 to 18:1. The gas temperatures that result from these ratios are unacceptable due to the thermal limitations of the turbine and turbine blade materials. It is therefore essential to cool the gases to an acceptable level. Introducing additional air into the gas stream after combustion has taken place achieves this, thereby increasing the overall air/fuel ratio to between 45:1 and 130:1.

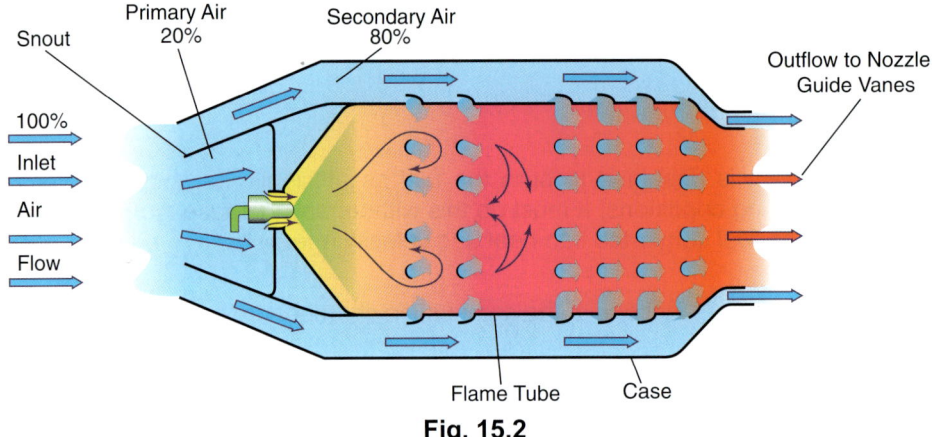

Fig. 15.2

Air leaving the compressor enters the combustion system where approximately 80% is directed into the space between the flame tube and the surrounding air casing. The remaining 20%, termed primary air, enters the flame tube via the snout, as shown in figure 15.2. Twelve percent (12%) of the primary air passes through the swirl vanes that surround the fuel burner into the primary zone, the remaining 8% of the primary air enters the primary zone via a perforated flare that stretches across the flame tube and supports the swirl vanes. Shown in Figure 15.3, this is sometimes termed a **colander**. The function of the swirl vanes is to impart a swirl to the air creating a vortex while simultaneously reducing axial velocity of the air to match the relatively slow burning rate of kerosene.

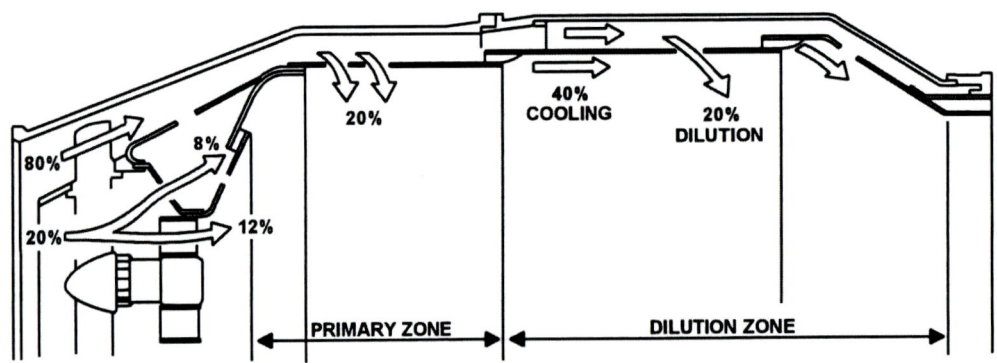

Fig. 15.3

An additional 20% of the secondary airflow enters the primary zone via secondary air holes. The interaction of these two airflows creates a region of low velocity recirculation that stabilises and anchors the flame, thus ensuring complete combustion and initiating cooling. Gas temperature during combustion is approximately 1800°C to 2000°C, which is excessive for entry to the turbine section. The remaining 60% of the secondary air is for cooling and dilution, of which approximately 40% cools the flame tube. The remaining 20% enters the dilution zone to dilute. This cools the gases to an acceptable level before they enter the turbine. Typical values of gas temperature entering the turbine are between 1000°C to 1500°C. Figure 15.4 shows an actual flame tube.

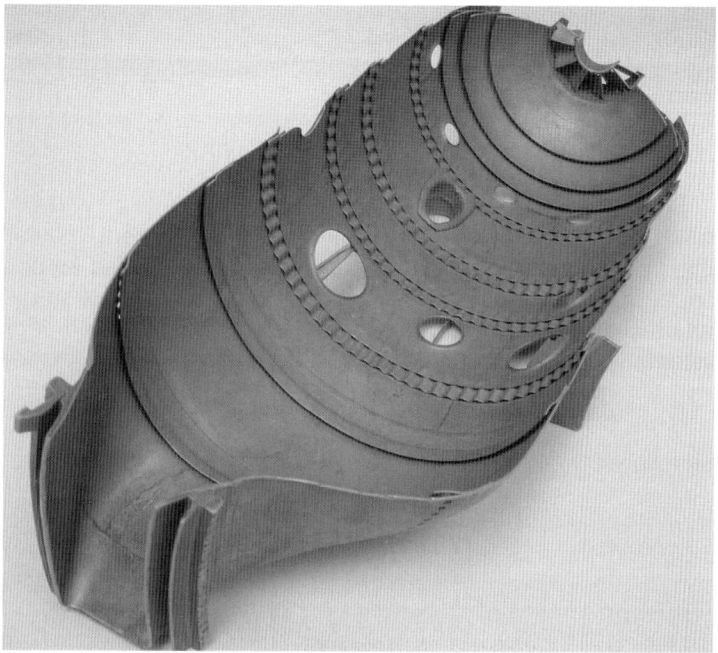

Fig. 15.4

Primary air is the portion of the compressor output air used for the actual combustion of fuel, usually 20% to 25%. Secondary air is the portion of the compressor output air used for cooling combustion gases and engine parts.

An igniter plug creates an electrical spark to ignite the air/fuel mixture and once ignited, the flame is continuous. There are two igniters per combustion system. Their operation is described later under ignition.

TYPE OF SYSTEM

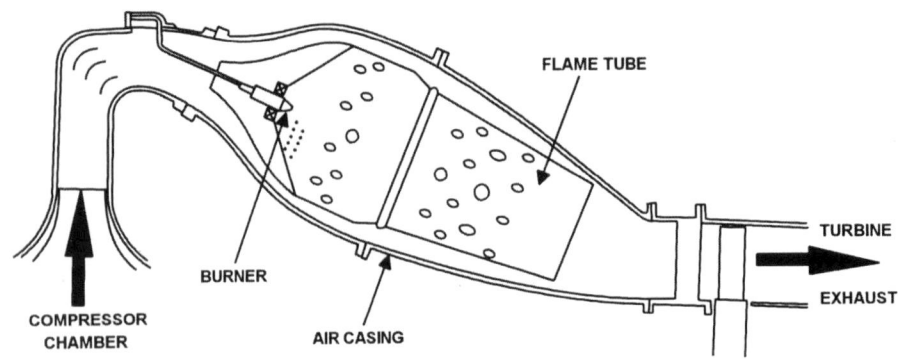

Fig. 15.5

There are three basic systems and, while they look somewhat different, they all consist of a flame tube within which the flame must be contained. It has a number of holes for complete combustion, cooling and dilution, an air casing to carry secondary, cooling, and dilution air, and a fuel nozzle to provide atomised or vaporised fuel for efficient combustion. Figure 15.5 depicts the earlier form of combustion systems as used on centrifugal compressor engines pioneered by Sir Frank Whittle.

The systems are as follows:

  ➤ Multiple
  ➤ Tubo-Annular
  ➤ Annular

MULTIPLE CHAMBER

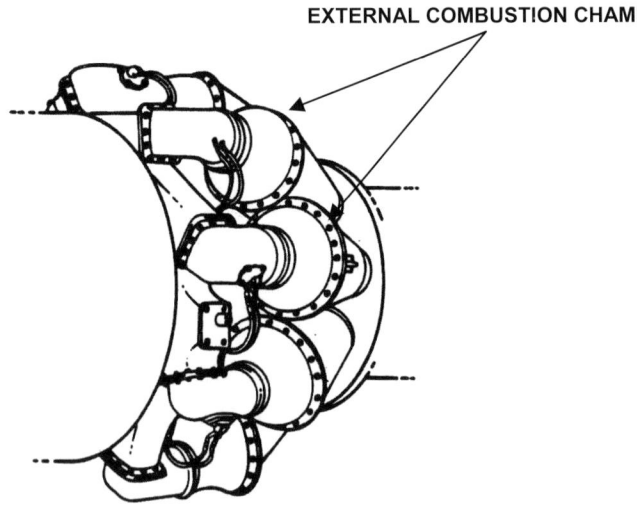

Fig. 15.6

This system has a number of inter-connected chambers contained within their own air casings mounted externally in a circle around the centre line of the engine. Except for fuel drains and igniters, the chambers are identical in construction.

The inter-connectors are required for flame propagation during ignition and for even pressure distribution throughout the combustion section.

TUBO-ANNULAR OR CANNULAR

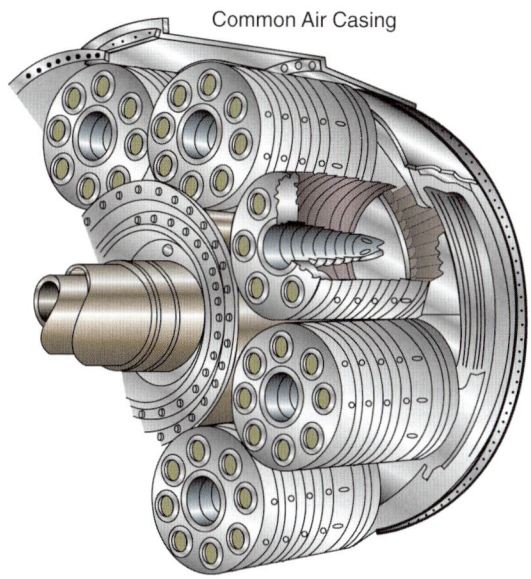

Common Air Casing

Fig. 15.7

The tubo-annular is the "half-way house" in design between the separate chambers of the multiple and single chamber of the annular type. It uses an annular air casing around the engine centre line and individual inter-connected flame tubes, fitted within the air casing.

ANNULAR

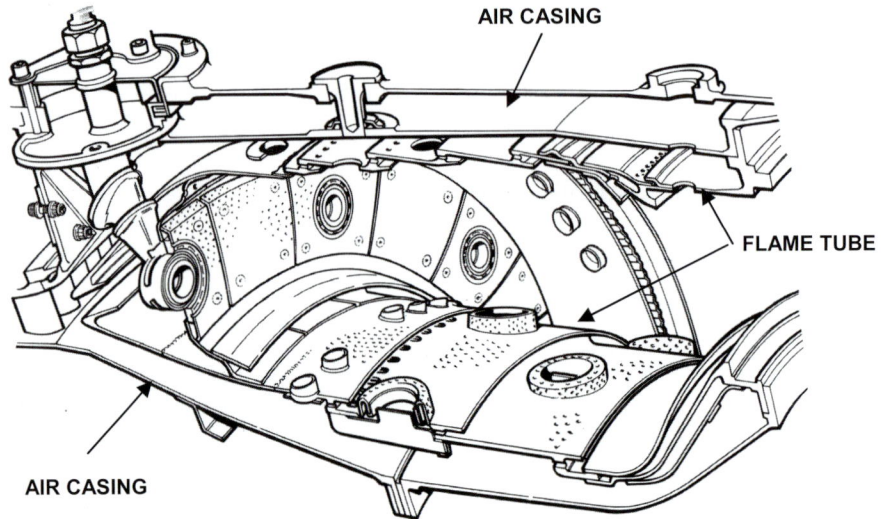

Fig. 15.8

The annular combustion chamber system has a single chamber that surrounds the engine, obviating the need for inter-connectors. Annular inner and outer air casings form a tunnel around the centre line of the engine. The flame tube is located in the space between the inner and outer air casing. Some designs incorporate a reverse flow chamber, where the air ducts to the rear via the air casing, travelling forward before being introduced into the flame tube and being directed rearward as normal.

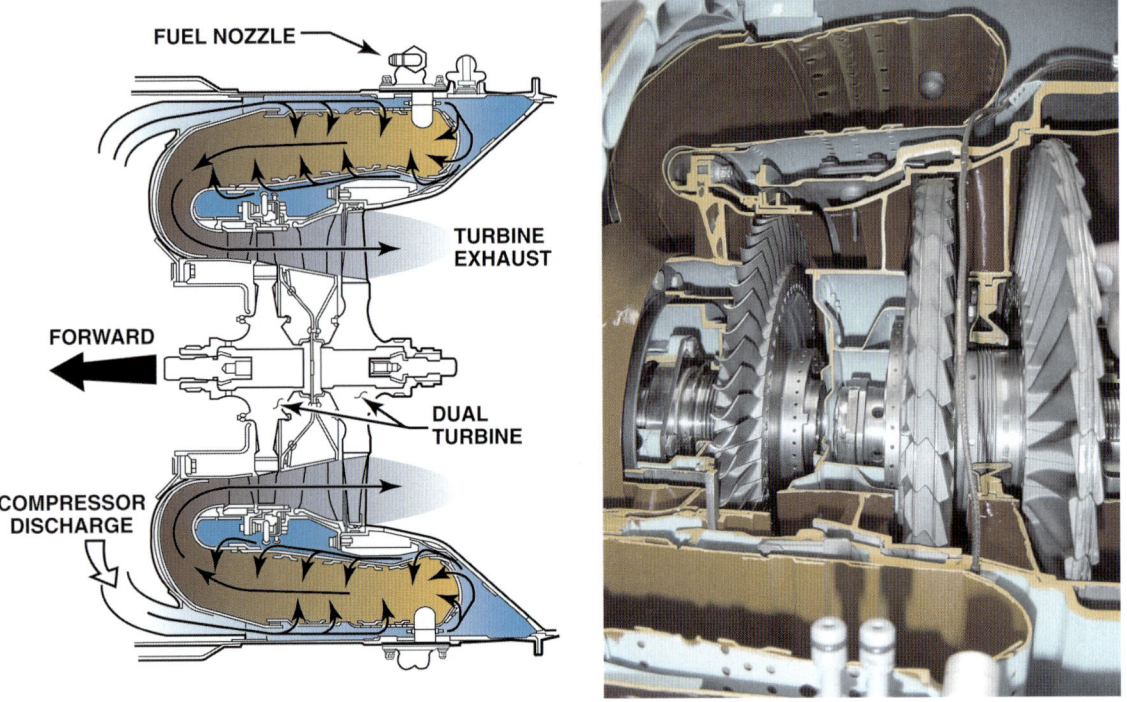

Fig. 15.9

Figure 15.9 shows a reverse flow annular combustion chamber. This system is frequently used in small turbo-prop engines as it allows the engine to be lighter and shorter, while reversing the flow allows the combustion chamber to preheat the compressor discharge air. These factors help compensate for the loss of efficiency resulting from the combustion chamber exiting forward.

The three designs can be summed up simply by saying that earlier engines contained the multiple types and, although somewhat bulky, it was simple to dismantle and service. The tubo-annular has some of the advantages of the multiple, but is more compact, with a smooth exterior and reduced weight. It appeared on later engines. The annular system, as used on the latest engines, provides a much more compact system and, for the same power output and mass flow, a much shorter one saving approximately 25% of length.

The reverse flow annular system has the advantage of shortening the overall length; however, it does create an engine with a larger frontal area, and suffers from efficiency losses. The reverse flow system is normally for use with a centrifugal compressor that provides the initial turning of the airflow.

FUEL NOZZLES

Because of the low volatility of kerosene, efficient combustion can take place only if the fuel is atomised or vaporised. Most gas turbines use the principle of atomisation using swirl atomisers to break the fuel down into fine particles.

The function of the nozzle is to inject the atomised fuel into the combustion chamber. The most common types of nozzle are:

> The Vaporiser
> The Atomiser

VAPORISER TYPE

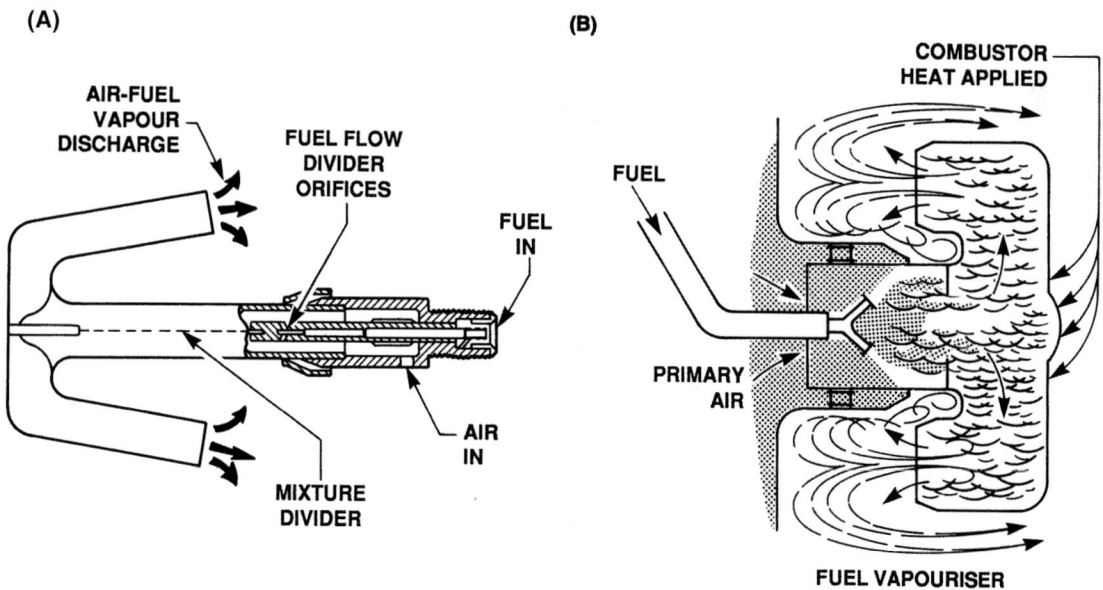

Fig. 15.10

In the vaporising method, fuel sprays from a feed nozzle into vaporising tubes located inside the flame tube. These vaporising tubes turn the fuel through 180° and are heated by the combustion process, thus vaporising the fuel before it passes into the flame tube. During starting, an additional method of heating is generally used.

ATOMISING TYPE

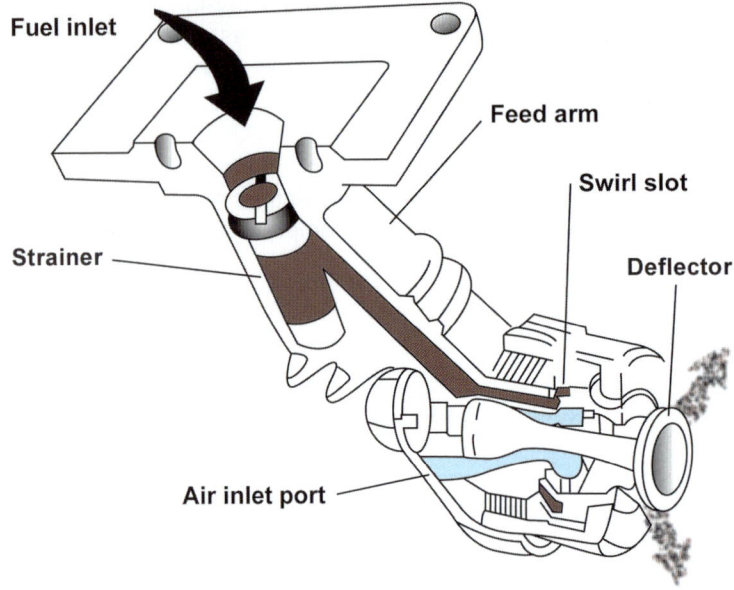

Spray nozzle

Fig. 15.11

A swirl atomiser consists of a small cylindrical or conical cavity into which a number of streams of fuel enter through tangential holes. The direction of fuel entry into the cavity creates a vortex and the swirling fuel leaves in atomised form from a single orifice.

SIMPLEX

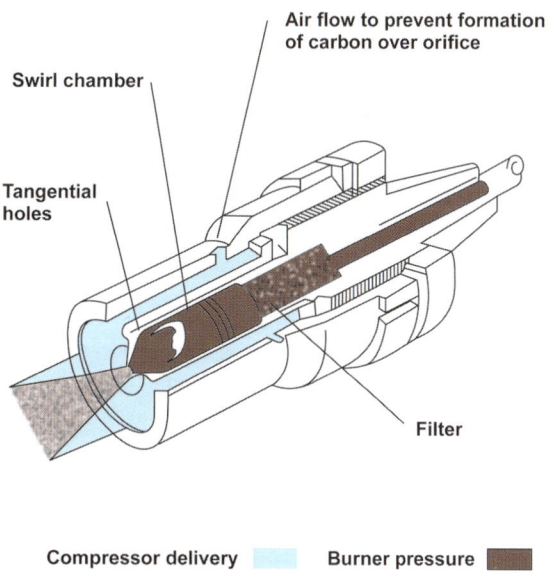

Fig. 15.12

This early nozzle consists of a simple swirl chamber and a fixed-area atomising orifice. This type of nozzle, whilst it provided good atomisation at high fuel pressures, was unsatisfactory at low fuel pressures and at low engine rpm, especially at high altitudes.

DUPLE AND DUPLEX

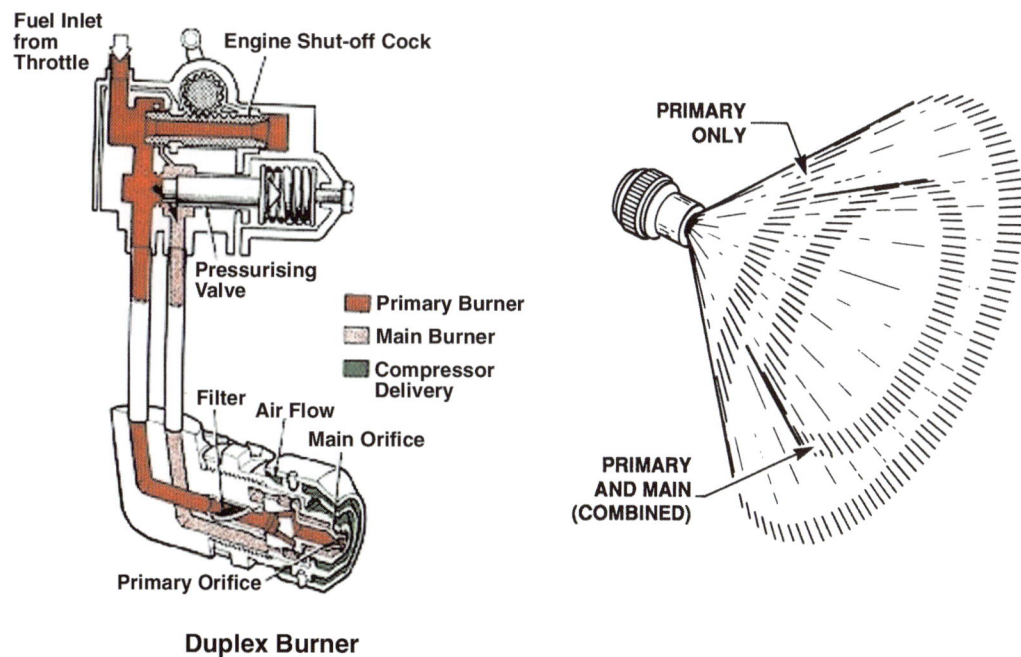

Fig. 15.13

These nozzles have two sets of atomisers, one for low flows and both combined for high flows. Two manifolds, the primary and the main, supply fuel to the nozzles. Each nozzle has two swirl chambers, one for the primary flow and the other for the main flow. At idling speeds and high altitudes, a pressurising valve permits fuel to pass only to the primary orifice. At higher flows and therefore higher pressures, the pressurising valve opens to allow fuel to pass to the main orifice as well, thus providing a combined flow to the nozzle. In this way, this type of nozzle can provide effective atomisation over the whole operating range of the engine.

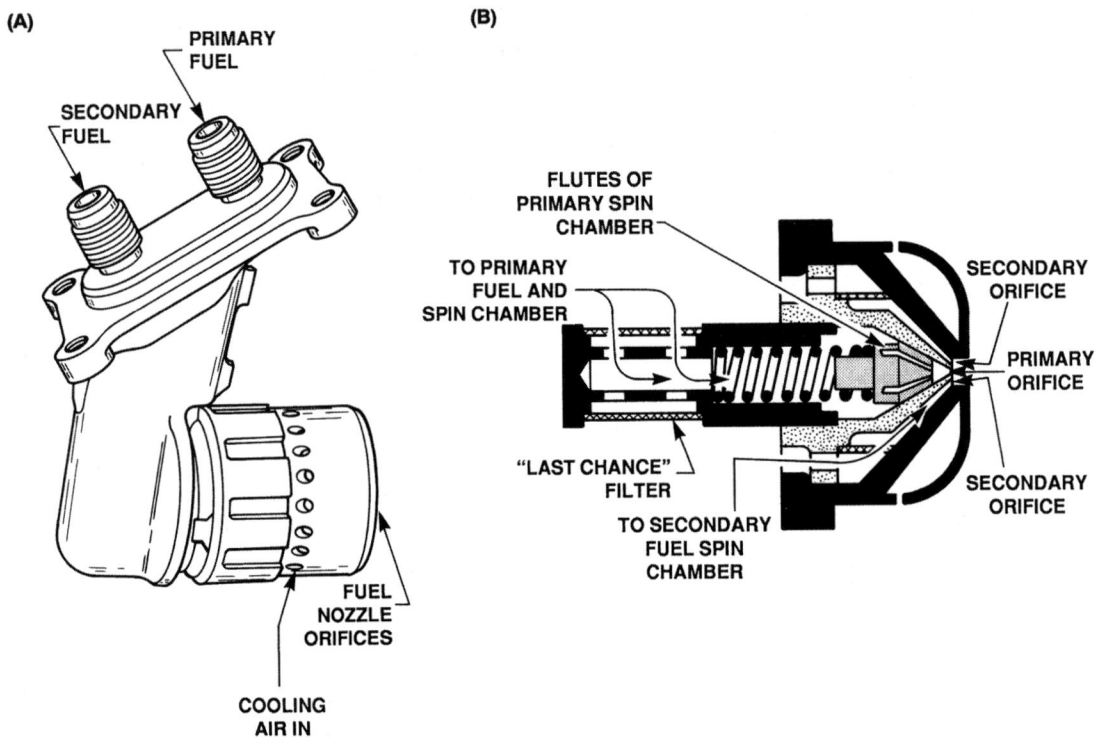

Fig. 15.14

Referring to figure 15.13, the main flow has a narrower cone angle to ensure that the highest temperatures remain away from the flame tube walls. Other duplex systems utilise two separate fuel lines to the nozzle assembly as shown in figure 15.14. The duplex systems require high fuel pressures (800 to 1500 psi) to ensure that the fuel is atomised and obtains the correct flow pattern. This high pressure has the advantage of providing the duplex designs good resistance to fowling due to contaminants within the fuel or carbon deposits.

SPILL-TYPE

This nozzle has a passage from the swirl chamber for spilling fuel away. This makes it possible to supply fuel to the swirl chamber at a high pressure all the time. As fuel demand decreases with altitude or engine speed reduction, more fuel spills away, resulting in less fuel passing through the atomising orifice.

The advantage of this type of nozzle is that the constant use of high pressure fuel means that even at the extremely low fuel flows that occur at high altitude, there is satisfactory swirl providing constant and efficient fuel atomisation.

ROTARY ATOMISER

A rotary atomiser feeds fuel into the centre of the main shaft, then along to drillings in the shaft in the combustion area. The fuel is atomised by the centrifugal action of the shaft.

SPRAY NOZZLE

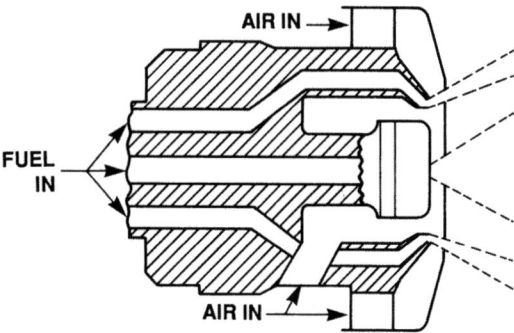

Fig. 15.15

Primary air is introduced with the injected fuel. Aerating the spray in this manner achieves greater burning efficiency, therefore reducing both carbon formation and exhaust smoke. The comparatively low pressures required for atomisation also makes it possible to use the lighter gear-type pump.

Chapter 16
Gas Turbine Engine Turbines

INTRODUCTION

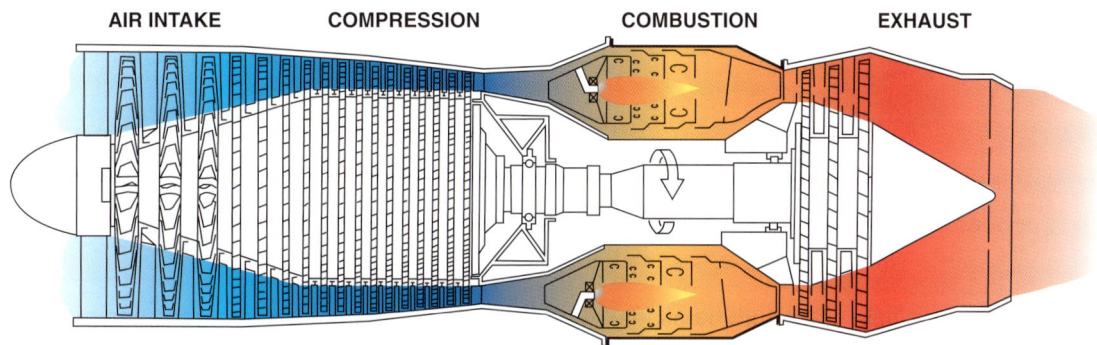

AIR INTAKE **COMPRESSION** **COMBUSTION** **EXHAUST**

Fig. 16.1

The purpose of the turbine section in a turbojet is to use some of the energy in the rapidly moving hot gases exiting the combustion section to turn a shaft to drive the compressor, or in the case of a turboprop to drive the propeller via a reduction gearbox. They may be single-shaft or multi-shaft (two or three shaft) arrangements.

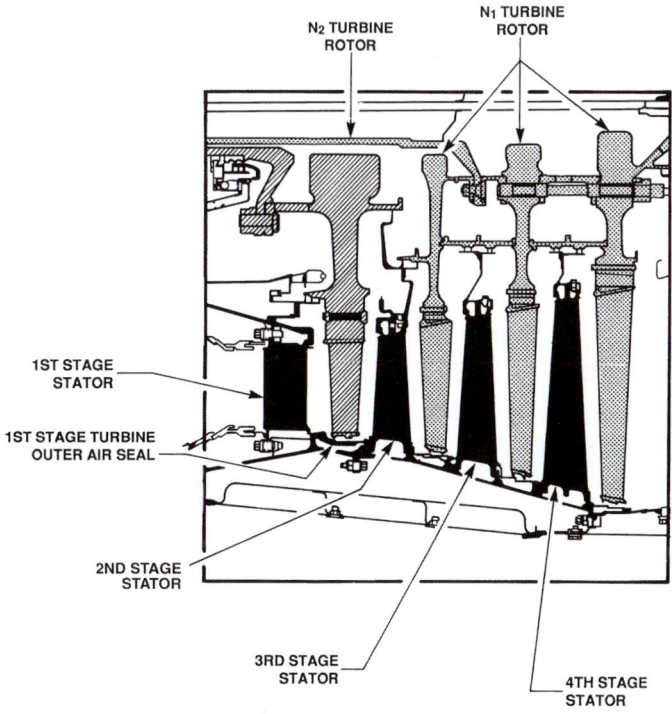

Fig. 16.2

The turbine assembly consists of a row of stator vanes, called nozzle guide vanes (NGVs), which are mounted in the engine casing. A row of loosely mounted turbine rotor blades follows to allow for thermal expansion on a disc or wheel. The most common type of rotor fixing is the fir tree root.

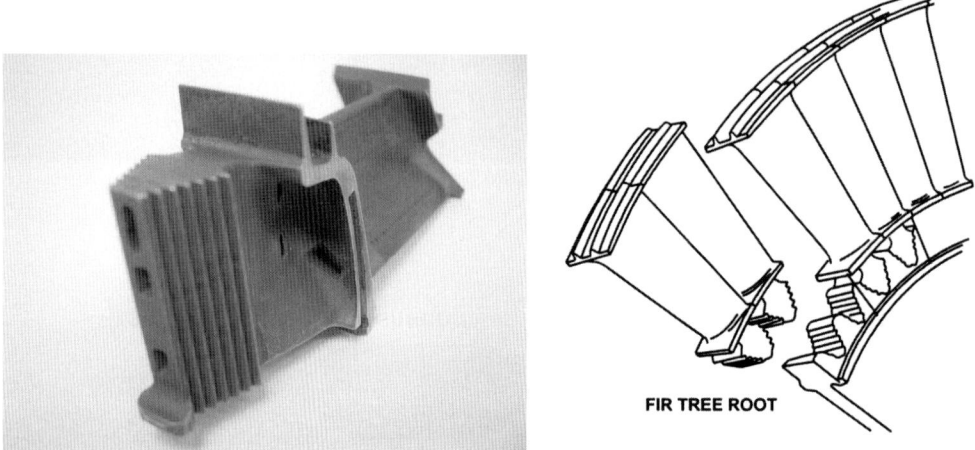

Fig. 16.3

This constitutes a single stage turbine assembly. Different numbers of turbine stages can be used in an engine, depending on the design requirements. Often, there are one or two stages of a high-pressure turbine. Possibly, three or more can make up a low-pressure turbine on a two-shaft arrangement. In the case of a triple-shaft, one or more stages make up an intermediate-pressure turbine. Figure 16.4 illustrates a two-shaft arrangement.

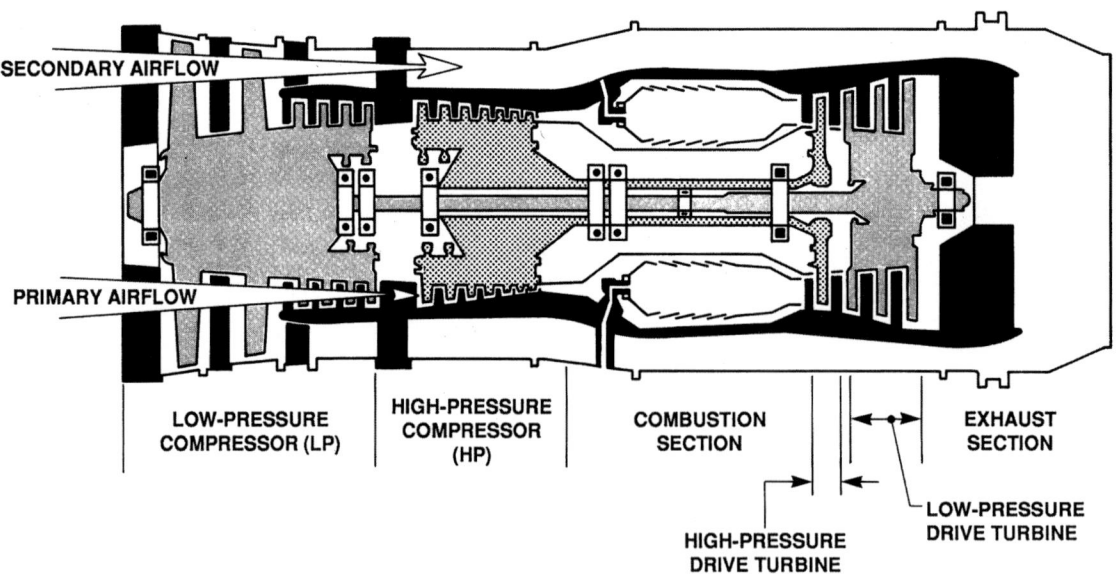

Fig. 16.4

The multi-shaft arrangement allows each turbine to run at its optimum speed since it is mechanically independent of the other turbine shafts. In the multi-shaft arrangements, the LP compressor, fan, or propeller connects to the LP turbine, which is the rearmost turbine and runs the slowest. In the case of a triple-shaft arrangement, the IP compressor connects to the IP turbine located in front of the LP turbine, and runs faster than the LP turbine. Finally, the HP compressor connects to the HP turbine, which is located immediately after the combustion section and in front of the IP/LP turbines and runs faster than the previous mentioned turbines.

TURBINE PRINCIPLES OF OPERATION

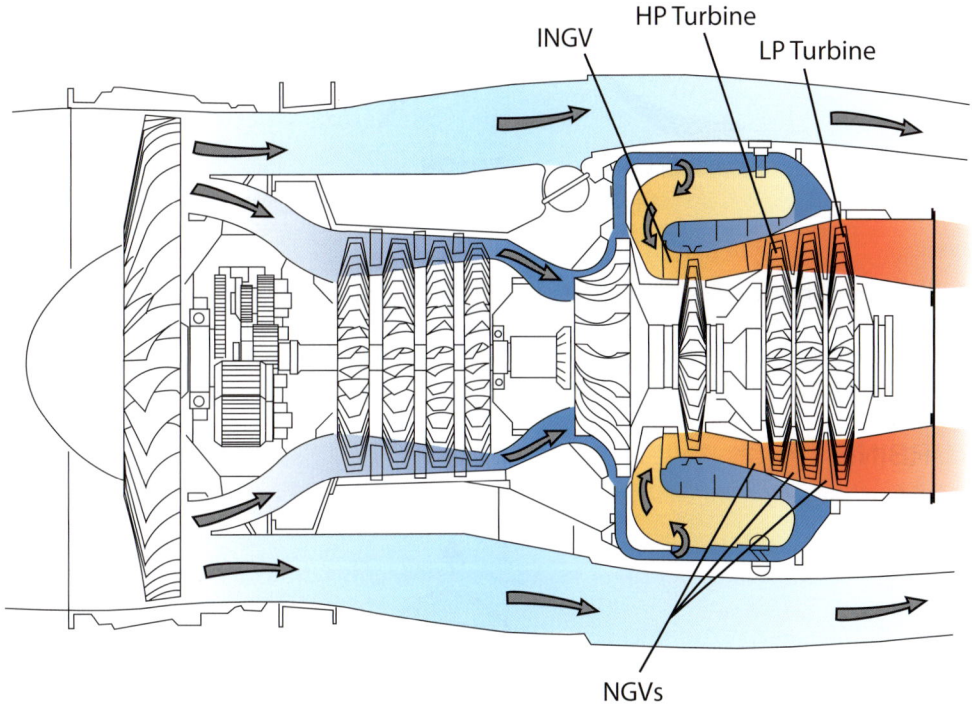

Fig. 16.5

The turbine section of an engine consists of rotors and stators, the rotors being the turbine itself and the stators being fixed blades that guide the gas flow onto the turbine blades at the correct angle. These are termed nozzle guide vanes (NGV). The first nozzle guide vane mounts immediately after the combustion chamber and is termed the inlet nozzle guide vane (INGV). It is through this guide vane that the gases reach their highest velocity. Figure 16.5 illustrates this.

Theoretically, there are three types of turbine blade that can extract power from the exhaust gases. These are reaction, impulse, and a combination of impulse and reaction.

REACTION TURBINE

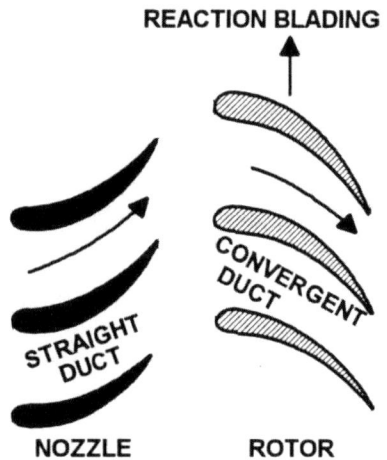

Fig. 16.6

In the case of a pure reaction turbine, the NGVs, by design, alter the gas flow direction without changing the pressure. As a result, they are parallel. The passageways between the blades of a reaction turbine are convergent and so accelerate the air. The reaction to this acceleration is felt in the opposite direction and so drives the turbine.

IMPULSE TURBINE

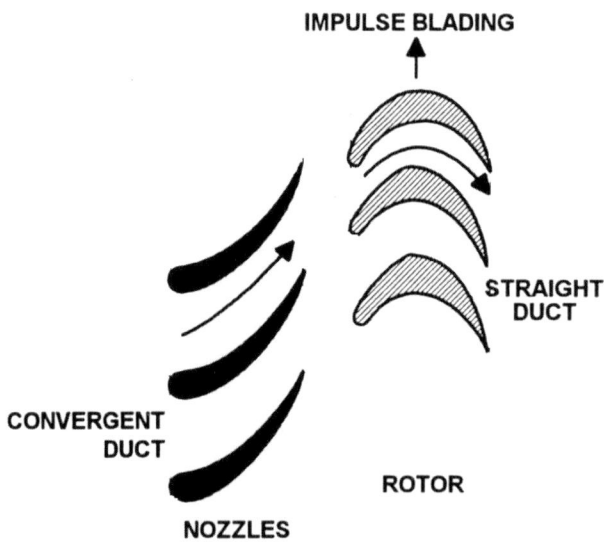

Fig. 16.7

A pure impulse turbine is designed to take full advantage of the high gas velocity from the convergent NGVs. The impulse force caused by the impact of the gas on the blades drives the turbine and the passageways between the blades are parallel.

Either the gases impinging on the blade or accelerating between the blades produces torque. Whilst it is possible to have a completely impulsive turbine (e.g. air starter motors), no blades are completely reaction based. Some degree of impulsive force always exists. Therefore, modern turbine engines normally use a combination blade that is both impulse and reaction.

IMPULSE/REACTION BLADES

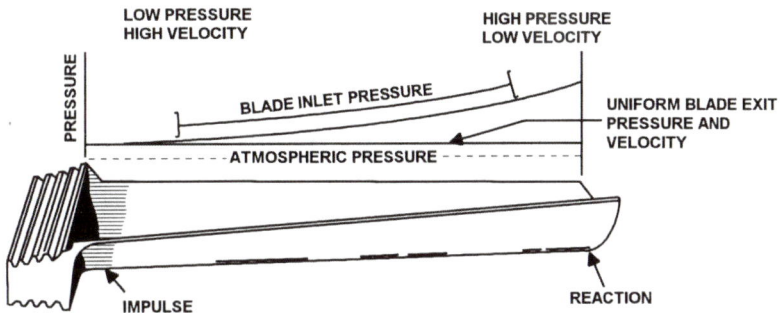

Fig. 16.8

As the gases travel through the passages of the nozzle guide vane and are directed onto the rotor blades to give the correct direction of rotation, the gas flow forms a vortex. In a vortex flow, the gas pressure increases and the velocity decreases toward the tip of the blade, whereas at the root of the blade the velocity is higher and the pressure is lower.

As this would result in the formation of an uneven load along the length of the blade, the design of the impulse/reaction blade serves to take advantage of the gas flow. Here, the blade is manufactured with an impulse shape at the blade root and a reaction shape at the blade tip. The impulse reaction is generally set to be 50/50 along the blade's length, as shown in figures 16.8 and 16.9.

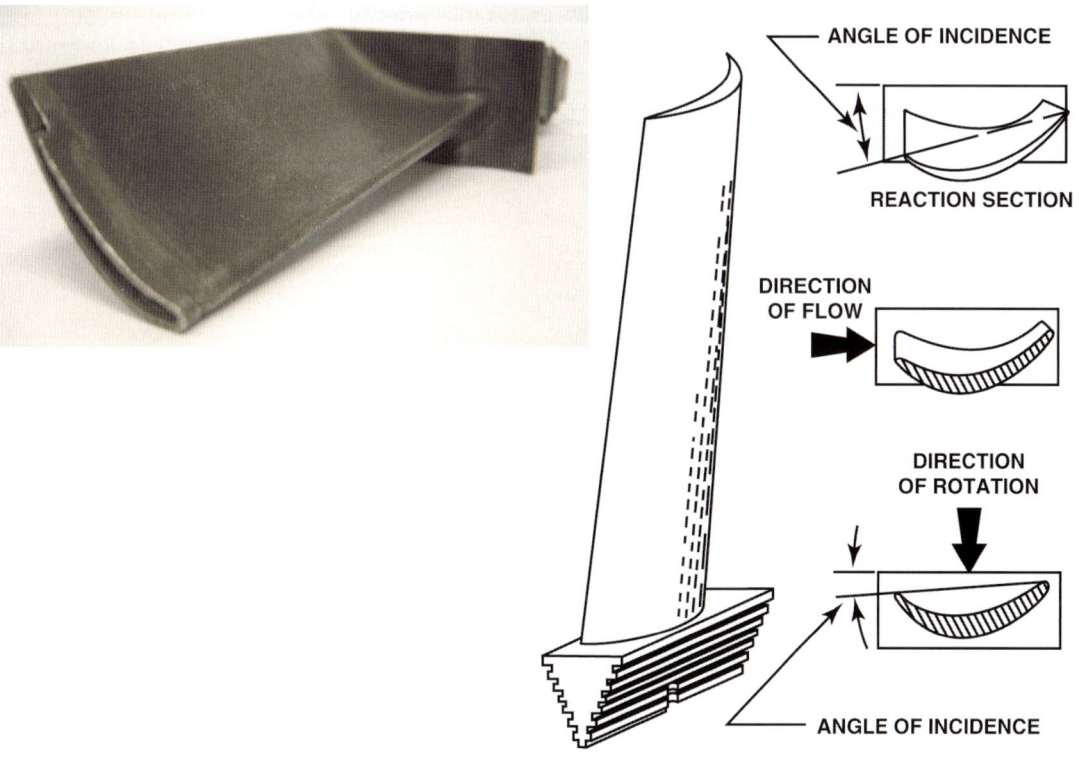

Fig. 16.9

Fig. 16.10

As the gas flows through the turbine section, its pressure and velocity decrease due to the work it does in rotating the turbine discs. In addition, this increases the need for each subsequent turbine disc to have longer blades to obtain the maximum usable energy. The gas having passed from the back of one turbine receives a swirl that is trued up by the following NGV prior to its entry to the next turbine. In the case of impulse reaction turbines, the NGV forms convergent ducts toward the centre and parallel ducts toward the outside. A cross section from the turbine disc to the housing forms a divergent gas flow annulus, to account for expansion. See figure 16.10.

TURBINE COOLING

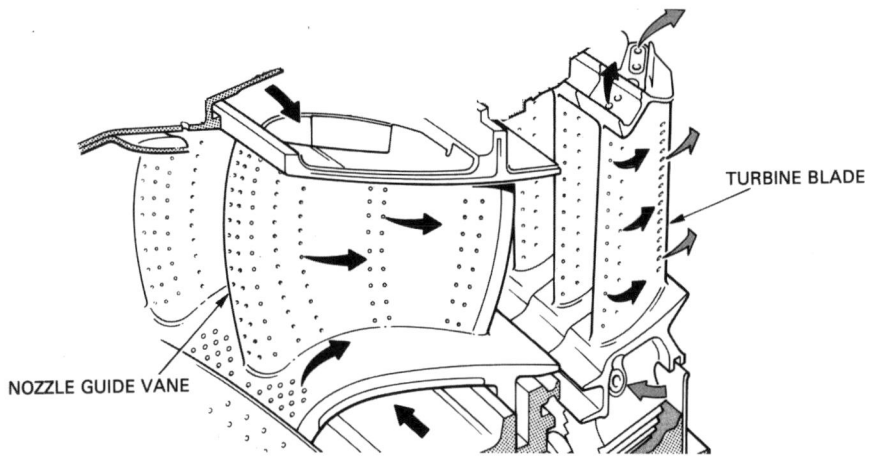

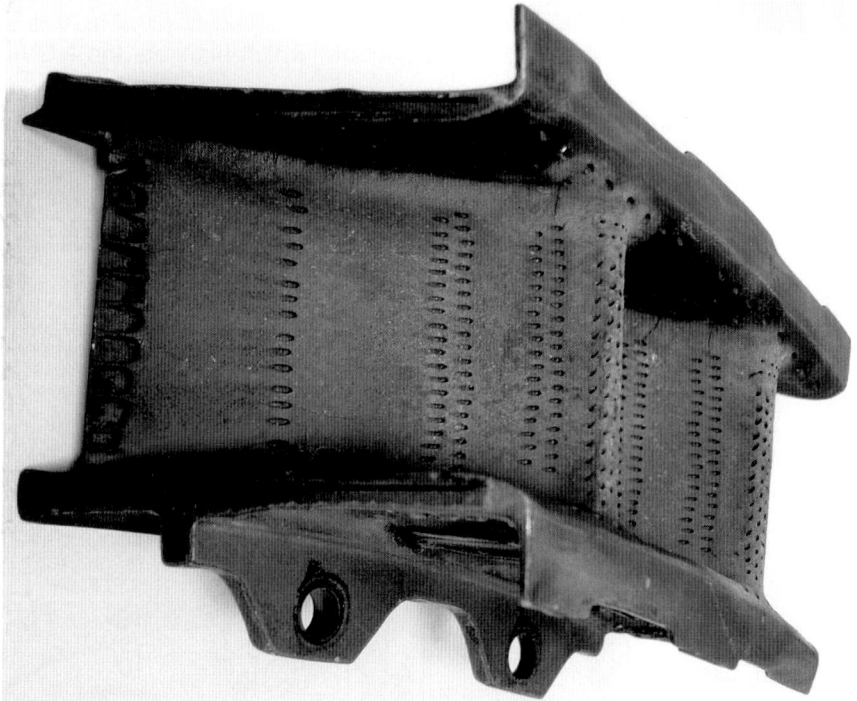

Fig. 16.11

Improved blade cooling has achieved a further advance in temperature increase. A percentage of the mass flow passes through longitudinal holes, cavities, or tubes formed in the blades and NGVs, which reduces the blade surface temperature by convection. Air can also impinge along the surface of both the NGVs and blades.

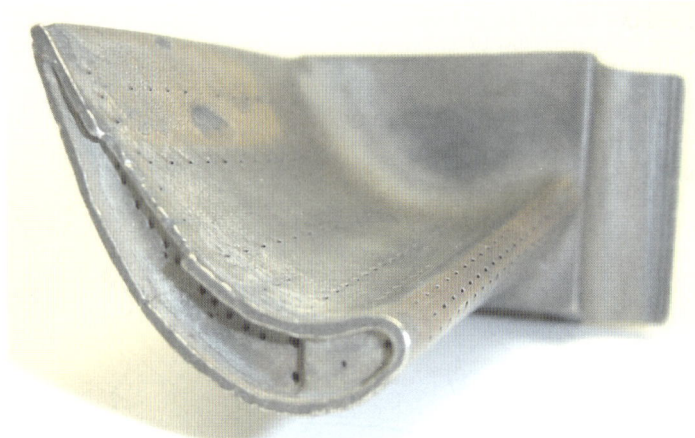

Fig. 16.12

Modern NGVs and blades cool via a combination of a complex internal convection and film cooling of the surface. Film cooling is a result of the ejection of a cool jet of air into the boundary layer of the NGVs and blades creating a film of cool air, which blankets the NGVs and blades from the hot gas.

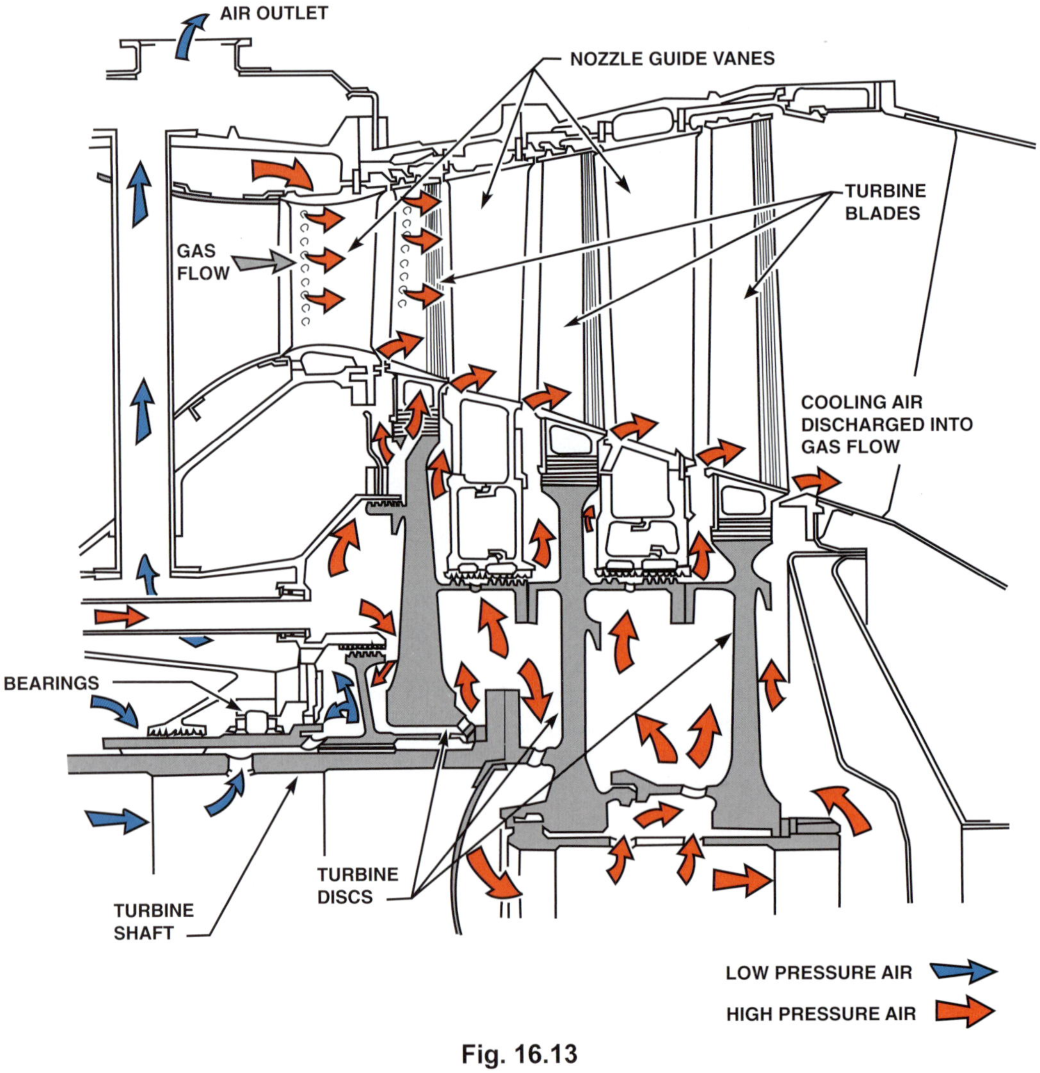

Fig. 16.13

The cooling air in this case may be taken from different stages of the compressor in order to have a graduated cooling of the hottest NGVs and blades (i.e. high-pressure turbine) to prevent thermal shock. This allows the inlet temperature to increase without affecting the material. Note that as the gas transits the turbine the temperature decreases, therefore reducing the cooling requirements and allowing employment of less sophisticated methods. Figure 16.13 illustrates a typical high pressure NGV and turbine blade cooling system.

EXHAUST GAS TEMPERATURE

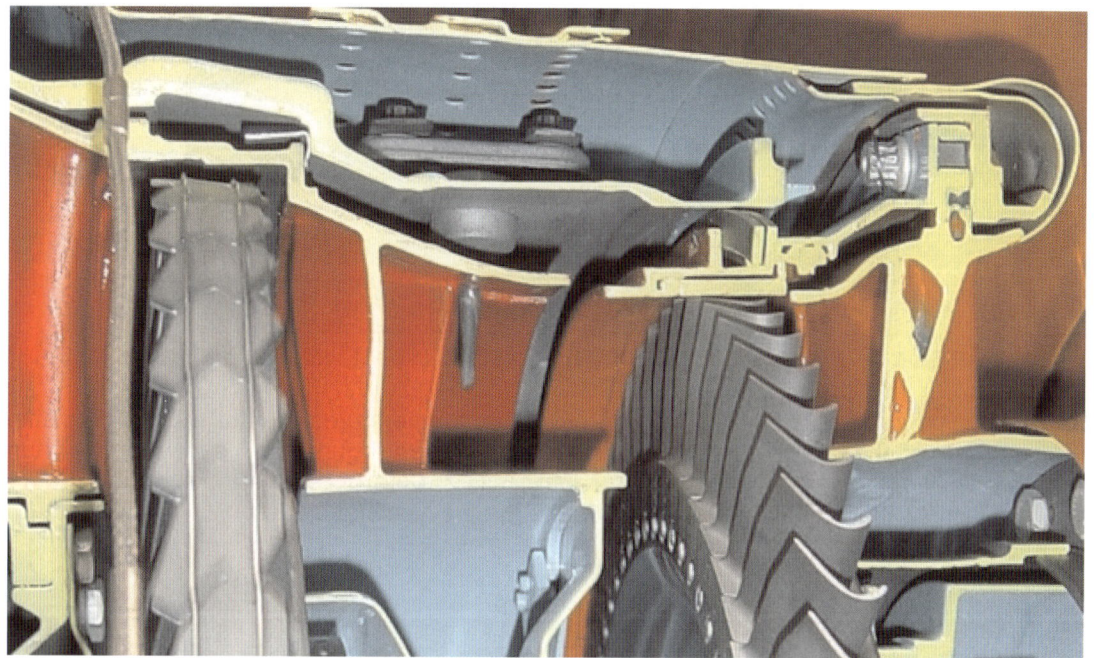

Fig. 16.14

In order to monitor turbine stress, the gas temperature leaving the turbine should be measured as close to the turbine entry point as possible. In figure16.14, the probe appears as a short, thin, grey tube protruding from the annulus between the two stages of turbine.

Over the years, the point at which the measurements occur has gradually moved toward that goal. Modern engines usually measure the gas temperature after the high pressure or low-pressure turbine.

The affect of engine acceleration is to increase the gas temperature. Take care that acceleration limits are not exceeded. Similarly, deceleration can cause overcooling resulting in turbine stress.

MATERIALS AND STRESSES

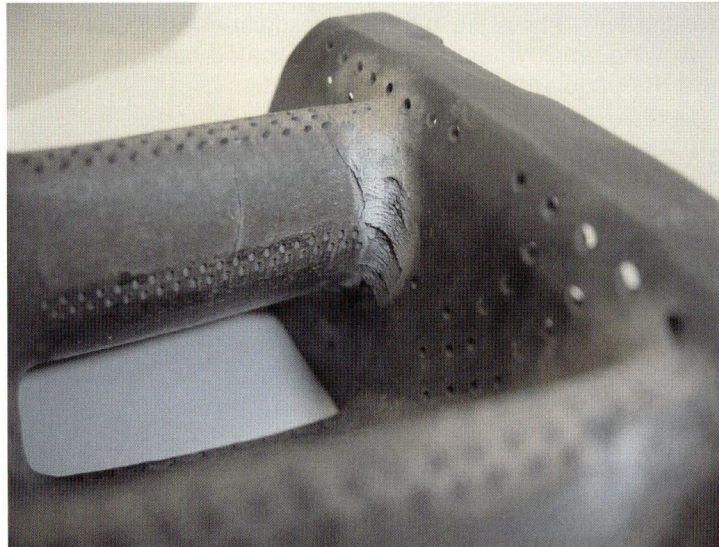

Fig16.15

One of the disadvantages of using higher turbine entry temperatures has always been the effects of temperature on the nozzle guide vanes and turbine blades. Figure 16.15 shows cracks caused by excessive temperature. The other factor that limits the life of the turbine blades is the high rotational velocity that imparts tensile stress to the turbine disc and blades.

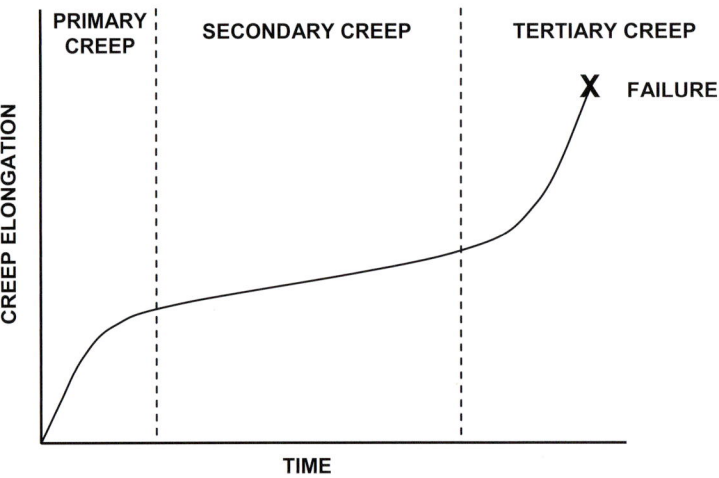

Fig. 16.16

The high stress makes it necessary to restrict the turbine entry temperature so the nozzle guide vanes, turbine discs, and blades can function for a satisfactory length of their working life without achieving the end of their useful creep life. Creep is the permanent elongation of the blades caused by temperature and time. There is a finite useful creep life limit before blade failure occurs. Figure 16.16 shows the three phases of creep. If the turbine entry temperature increases, an increase in material thickness and cooling airflow is required.

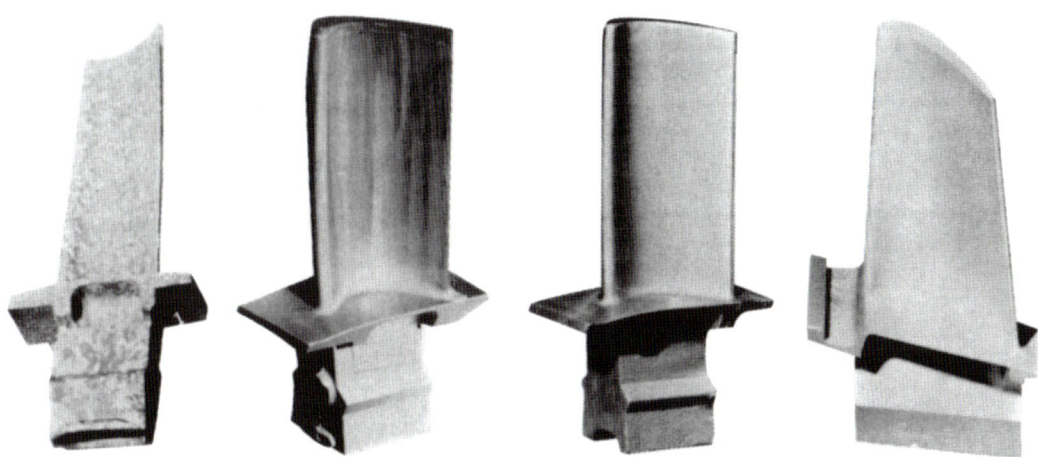

Fig. 16.17

Single crystal blades (pioneered by Rolls Royce) have longer in-service lives due to their method of manufacture. The blade cooling is strictly controlled so that the normal granular crystal structure caused by the uneven blade cooling due to the different thicknesses is prevented. Instead, a single grain or metal crystal forms; its boundaries being that of the blade. This removes the possible fracture lines; however, the blade still elongates over a period of time due to rotational velocity and turbine temperatures. As blades age they elongate and thin at the mid-span section, termed **hooking**, as the tip of the blade takes on a curved hook appearance.

SHROUDS

Fig. 16.18

Figure 16.18 shows the original design of turbine blades, which relied on the tip of the blades to rub against an abrasive strip so that they wore down as they grew in length. This meant that the gas flow near the tips was able to flow over the tip not around the blade.

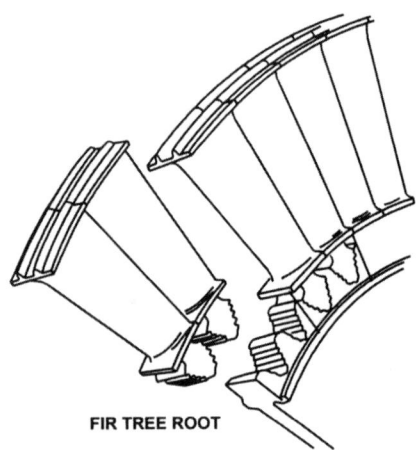

FIR TREE ROOT

Fig. 16.19

To overcome this, blade tips are manufactured with a shroud. When a disc is assembled, it takes the form of a wheel, the gas flows within the shrouded section and, for further sealing, the shroud may have one or two knife-edges that cut into an abradable strip to prevent leakage. Figure 16.19 illustrates this.

Modern engines that use the electronic FADEC system are able to use bleed air, taken from the compressor section, to control the temperature of the turbine housing. Controlling the housing's temperature controls the rate of expansion, enabling the gap between the blade shrouds and the casing to be kept to a minimum.

These developments have extended the life of the engines, and whereas the earlier engines required regular removal for overhaul or replacement of worn items, modern engines can remain on an airframe for years. To enable maintenance to occur in the most cost effective manner, the engine is divided into modules, each of which can be replaced without the engine being removed from it mountings.

Chapter 17
Gas Turbine Engine Jet Pipe

EXHAUST SYSTEM

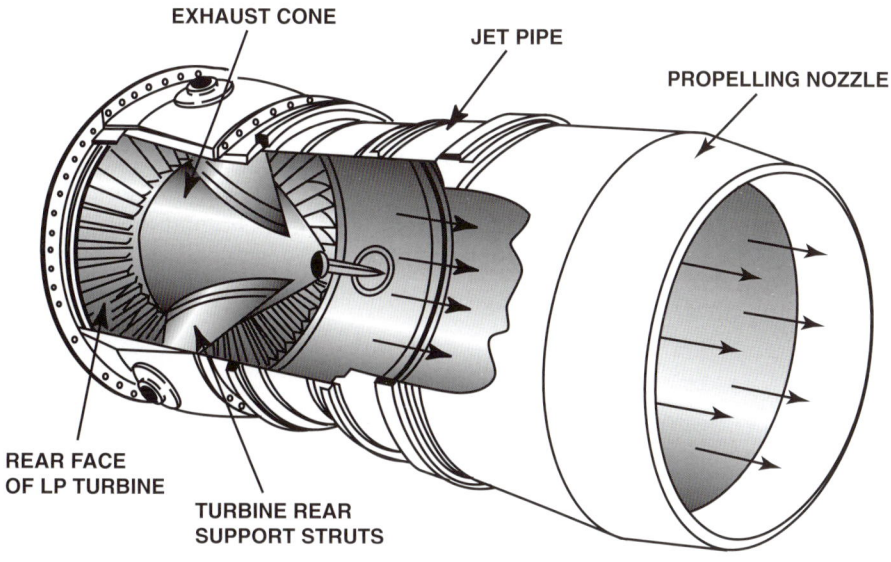

Fig. 17.1

The purpose of the exhaust system is to direct the exhaust gases to the atmosphere and to provide further acceleration of the exhaust gases, thus producing forward thrust. The exhaust system consists of:

- ➢ An exhaust cone
- ➢ A jet pipe
- ➢ A propelling nozzle

As the gas leaves the turbine, it travels at speeds between 750 to 1200 ft/second and at temperatures of approximately 550°C to 850°C or higher depending on the type of engine. At these speeds, high friction loss occurs. A danger of buffeting within the exhaust system also exists. As a result, the gas flow requires diffusing. The exhaust cone achieves this by forming a divergent section with the engine casing. Normally, the speed of the gas remains at approximately Mach 0.5 (950 ft/second). The cone also prevents the gas from flowing across the rear face of the turbine, with the support struts of the cone acting to remove any swirl in the gas.

The gas is then directed through the jet pipe, which is parallel and as short as possible to minimise frictional losses. The jet pipe connects to the propelling nozzle, which is convergent in shape to accelerate the gas as it ejects to the atmosphere. Additional thrust is obtainable under certain operating conditions where the exit speed reaches the local speed of sound for the gas temperature. In this condition, the nozzle is **choked**, and unless the temperature increases, the speed of the gas cannot increase further. The static pressure of the gas is greater than atmospheric pressure, and the pressure difference results in additional thrust, called **pressure thrust**. This type of thrust only occurs with the nozzle in the choked condition.

Note that due to the high velocity of the exhaust gas, a number of dangerous situations could arise. During ground handling, personnel can feel the effects of the exhaust gas far behind the engine exhaust pipe and therefore should exercise caution when near an aeroplane with engines running. In addition, a pilot should be aware of spectators, buildings, and other aeroplanes when manoeuvring on the ground. Operation of one aeroplane in the wake of another can have serious consequences. Loose objects thrown up by the leading aeroplane may be violently thrown against the following aeroplane or ingested by its engines.

VARIABLE AREA NOZZLES

Some engines use a variable-area exhaust nozzle. Using this type of nozzle creates an increase in the flow area through the nozzle, enabling easier starting at low rpm and temperature due to the reduction in turbine backpressure. A reduced area means increased thrust. The variations in nozzle area also enable low specific fuel consumption to be attained during some part of the engine operating range.

CONVERGENT/DIVERGENT NOZZLES

Convergent/divergent nozzles are used on some high pressure ratio engines to obtain the maximum conversion of energy in the combustion gases to kinetic energy and to increase thrust. In this arrangement, the convergent section exit now becomes the throat and the flared divergent section now becomes the exit.

On entering the convergent nozzle, static pressure decreases and the velocity of the gas increases. At the throat, the gas velocity is at the local speed of sound. On leaving the throat, the gas flows into the divergent section increasing in velocity toward the exit. The reaction to this velocity results in the inner wall of the nozzle being acted upon by a pressure force, thereby producing more thrust.

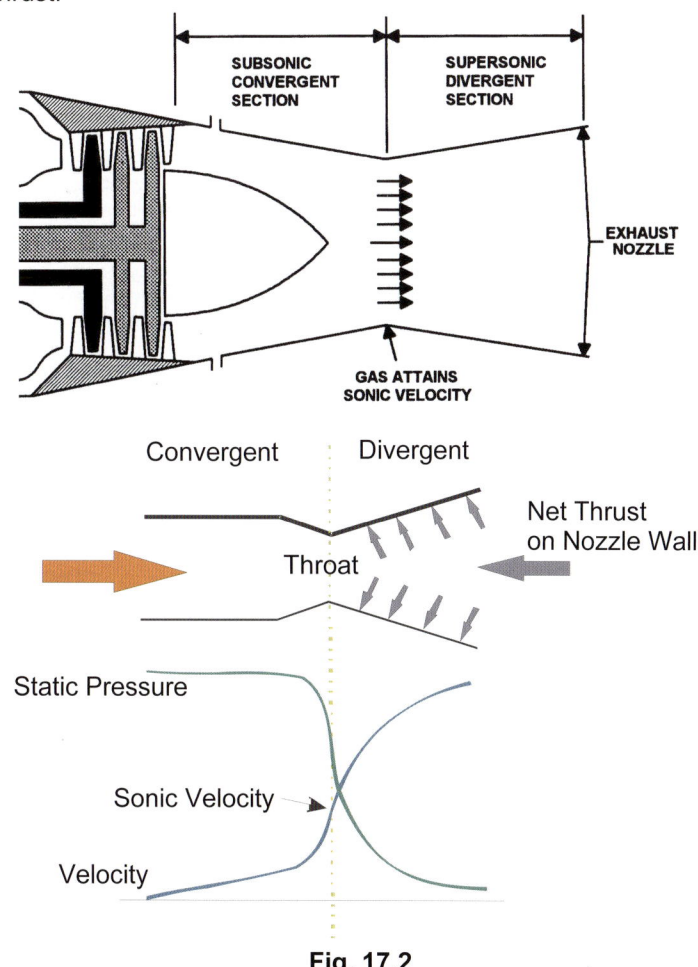

Fig. 17.2

OTHER DESIGNS

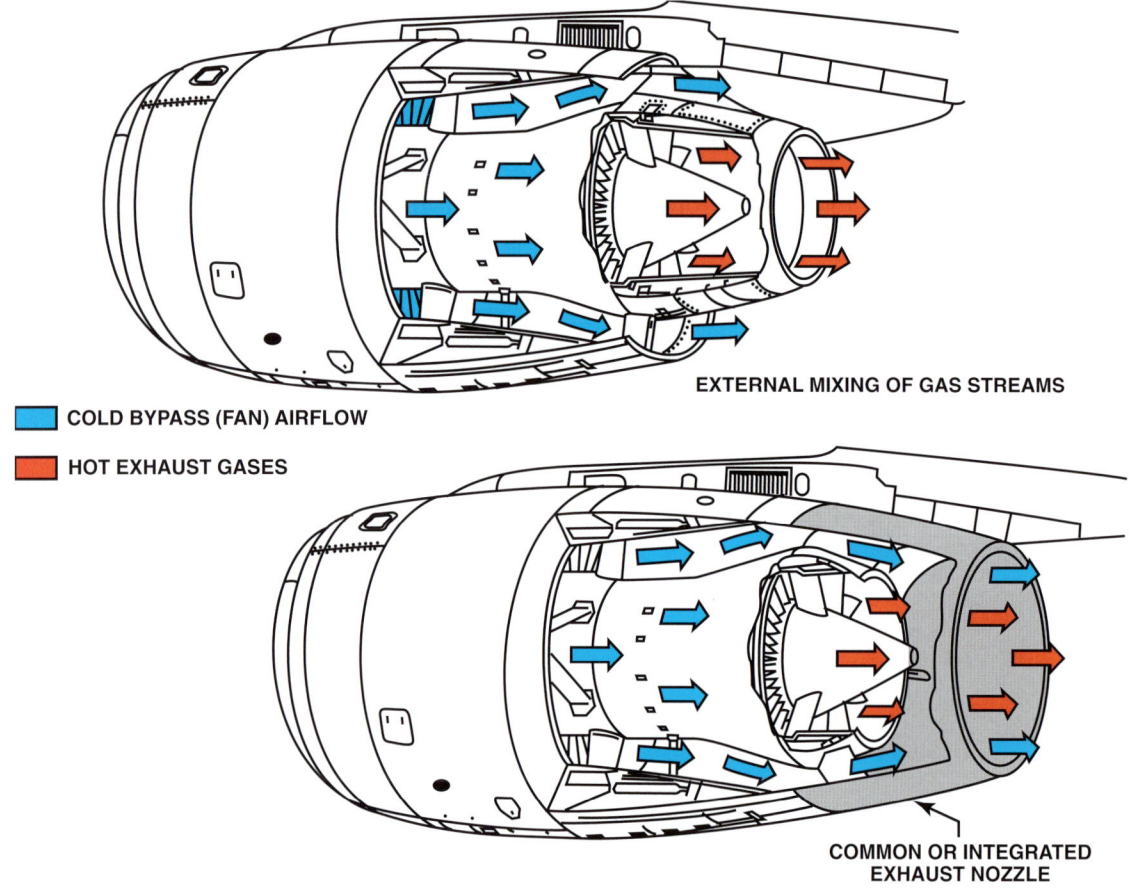

COLD BYPASS (FAN) AIRFLOW

HOT EXHAUST GASES

EXTERNAL MIXING OF GAS STREAMS

COMMON OR INTEGRATED EXHAUST NOZZLE

PARTIAL INTERNAL MIXING OF GAS STREAMS

Fig. 17.3

In the case of the low by-pass engine that has the cool by-pass airflow and the turbine hot gas to discharge into the atmosphere, the two flows combine via a mixing unit that allows by-pass air to enter into the hot exhaust gas in a way that ensures that the two streams mix thoroughly. High by-pass fan engines exhaust the hot gas and cold air streams separately via co-axial hot and cold nozzles designed to obtain maximum efficiency. On some installations, the two flows combine via an integrated or common nozzle that partially mixes them before they are exhausted to the atmosphere. This arrangement improves efficiency.

EXHAUST NOISE SUPPRESSION

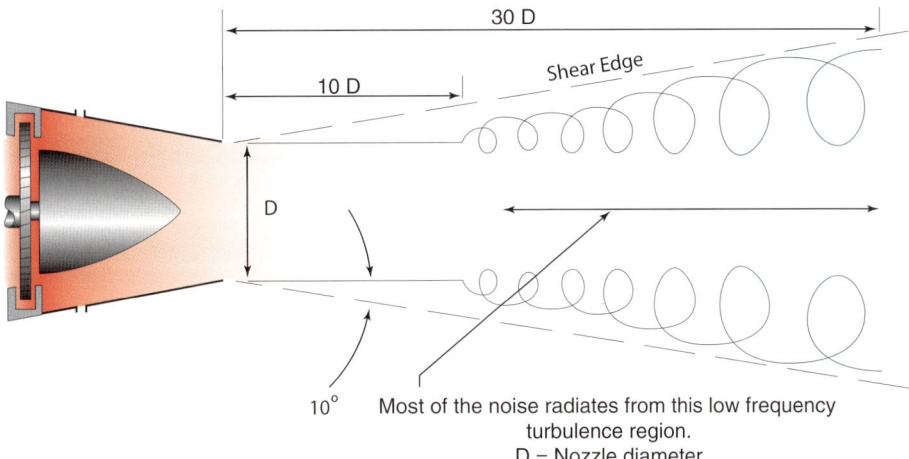

Fig. 17.4

The greatest source of noise from a gas turbine is due to the shearing action of the exhaust gases leaving the jet pipe and mixing with the atmosphere, producing turbulence. In general, the greater the jet velocity, the higher the noise level. As a result, one method of reducing noise is by decreasing the velocity, but this results in a reduction of thrust.

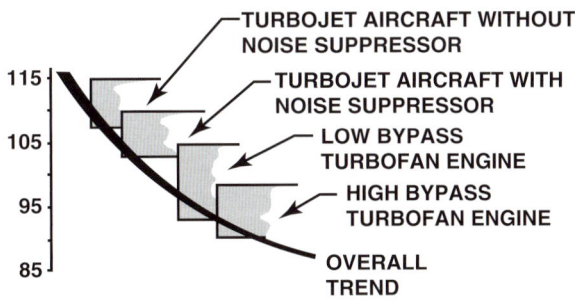

Fig. 17.5

Another cause of high noise level is the relative slowness of the exhaust gases mixing with the atmosphere. Therefore, using a device that speeds up the mixing results in a large reduction in noise. This device is a noise suppressor.

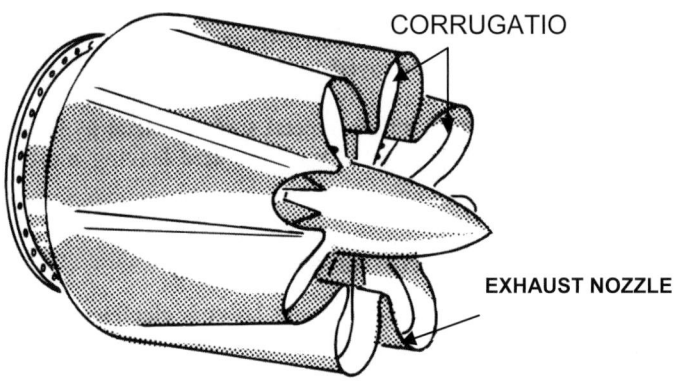

Fig. 17. 6

There are various designs of noise suppressors, but they all consist of a series of smaller nozzles built into the main propelling nozzle. They operate by allowing atmospheric air to be drawn through them to mix with the exhaust gas thereby increasing the area of contact between the jet efflux and the atmosphere. Suppressors that are more modern use deep corrugations or lobes to break up the main jet into a series of smaller jet streams to achieve the same effect, as illustrated by figure 17.6

The trend toward high by-pass ratio engines reduces the velocity of the hot gas, therefore reducing the shearing action.

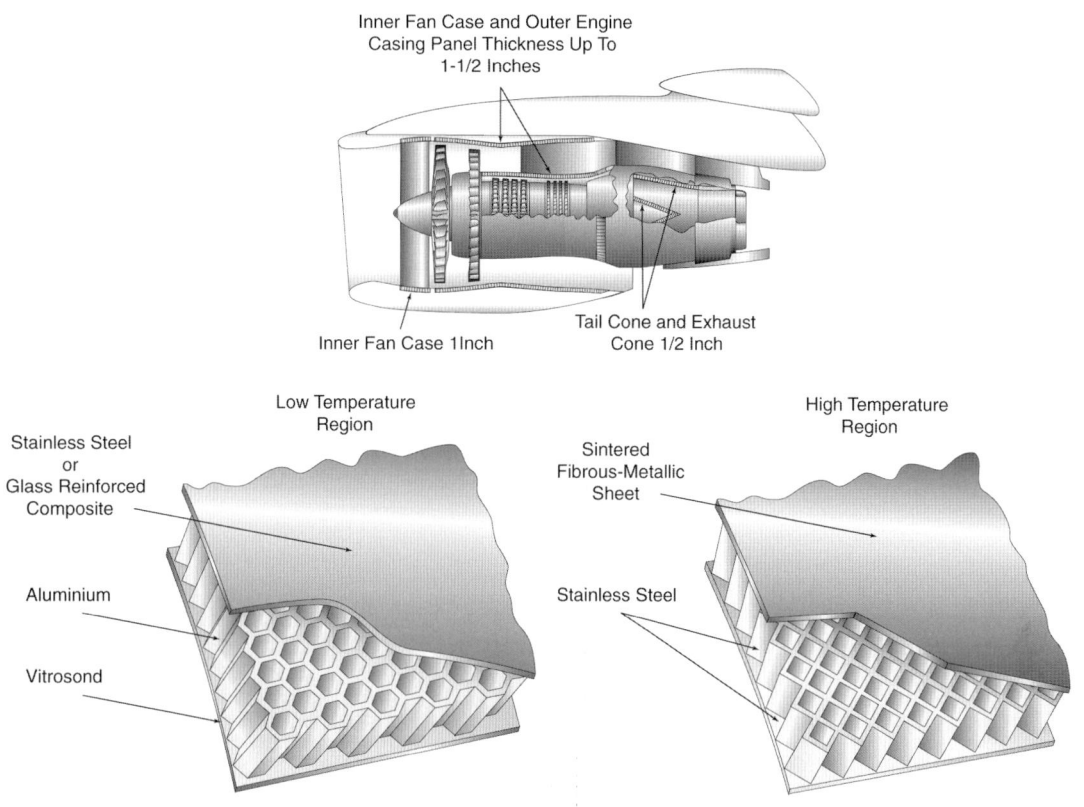

Fig. 17.7

All parts of an engine produce noise; as a result, noise absorbing material may be fitted in various locations to absorb the noise being produced (e.g. by lining the fan inlet and exhaust ducts).

Chapter 18
Gas Turbine Engine Reverse Thrust

INTRODUCTION

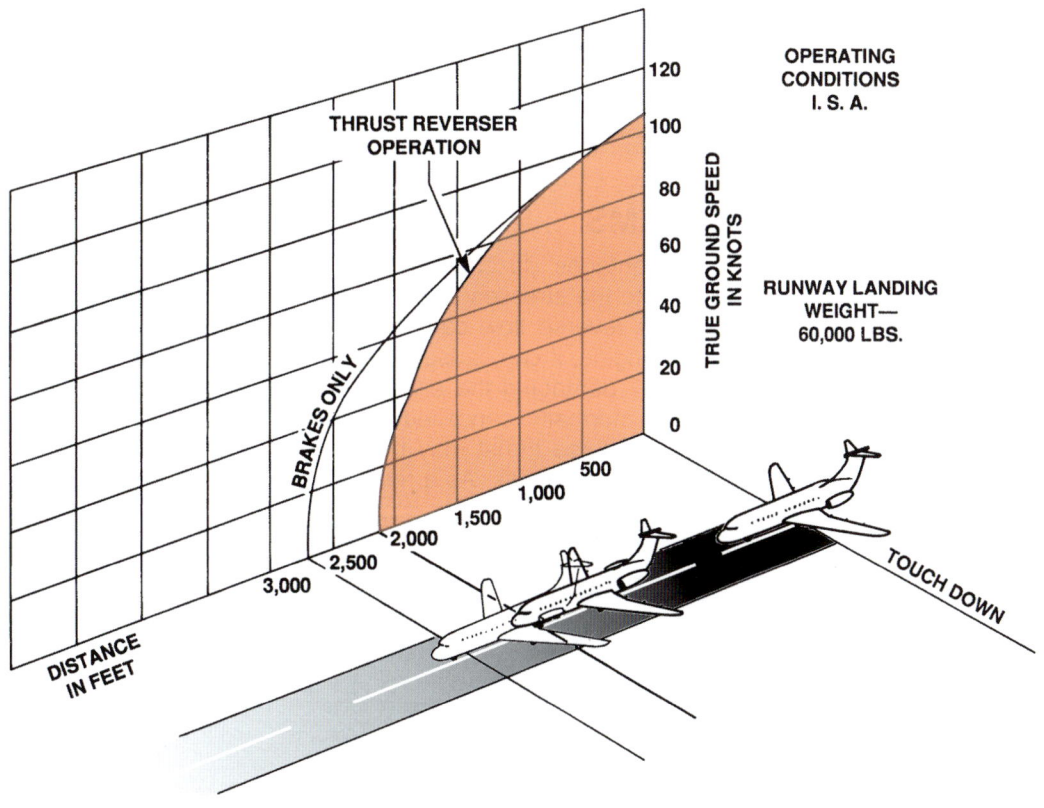

Fig. 18.1

Reverse thrust is a means of slowing down an aeroplane to reduce its landing run on both dry and contaminated runways. It also reduces high loads on the braking system, which decreases brake wear, tyre wear, and the associated risks of brake failure, fire, fade, and tyre burst. In an emergency, reverse thrust can also serve in an aborted take-off.

Reverse thrust systems reverse the direction of the hot exhaust gas for a turbojet and the cold air stream and/or the hot exhaust gas of a high bypass engine. A complete reversal of flow is not practical, mainly for aerodynamic reasons, so the angle of reverse is from approximately 45° up to 60°. Less thrust than normal is available in reverse thrust, and for a given rpm is approximately half that of forward thrust.

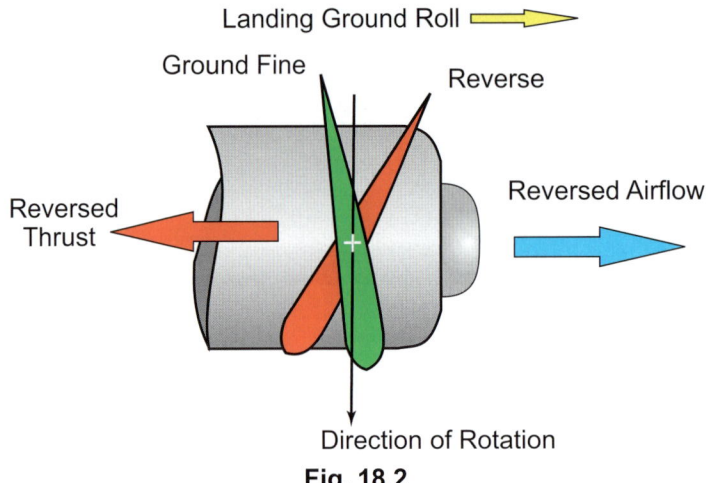

Fig. 18.2

In the case of a turboprop, the propeller pitch is reversed to a negative angle, which results in the air being accelerated forward.

OPERATIONAL PROBLEMS

Although the system has the advantages described above, there are some disadvantages associated with reverse thrust. Due to the stresses created during engine acceleration, it counts toward the ultimate engine life. The reverse flow can impinge on parts of the airframe. If the reverse flow is re-ingested into the engine, it can cause unstable engine operation. Debris kicked up by reverse flow may be ingested, damaging the engine. This situation is more likely at low speed; therefore, reverse thrust is normally cancelled at 60 kt. In some cases, especially turboprops, ground manoeuvring in reverse thrust may be allowed. Noise restrictions may limit the use of reverse thrust in certain circumstances. As a result, the pilot should ensure adherence to the correct procedures at all times.

REVERSE THRUST SYSTEMS

There are several methods for achieving reverse thrust. The most common are:

  ➢ Clamshell Doors
  ➢ External/Bucket Target Doors
  ➢ Blocker Doors

CLAMSHELL DOORS

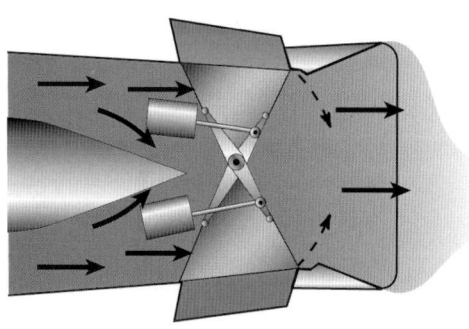 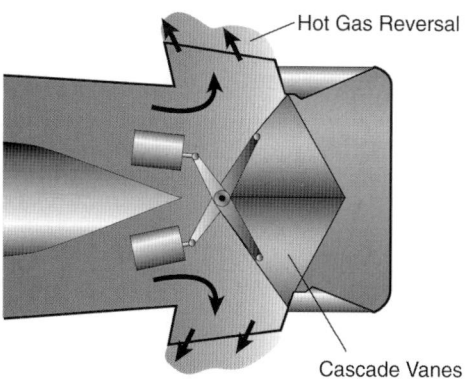

Fig. 18.3

These operate pneumatically, and do not affect normal engine operation as they form part of exhaust system whilst closed. When the pilot selects reverse thrust, the doors rotate into the hot gas stream, blocking the normal exit path of the gas stream, at the same time uncovering outlet ducts that contain cascade vanes. The cascade vanes direct the gas stream forward at the correct angle.

EXTERNAL/BUCKET TARGET DOORS

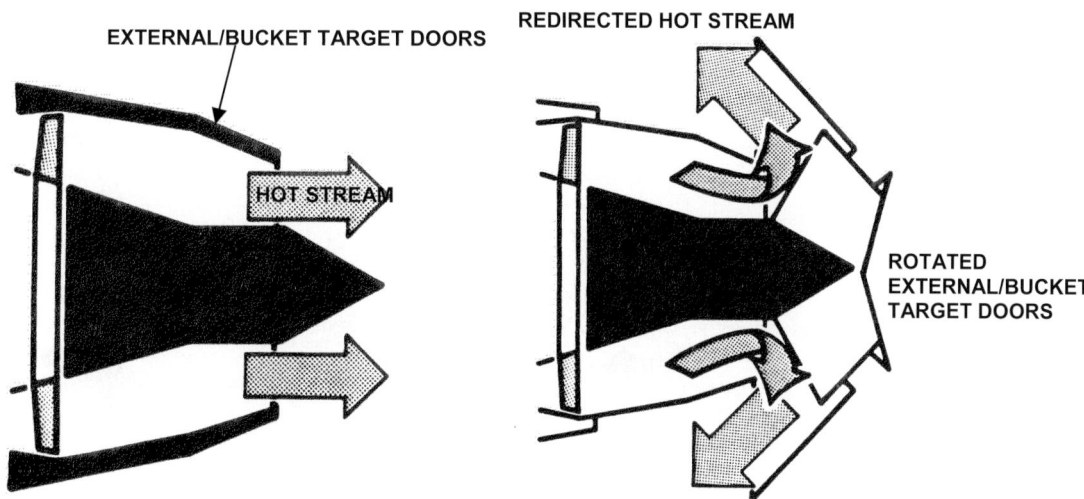

Fig. 18.4

External/bucket target doors operate hydraulically via a conventional push rod system. During normal engine operation they form the propelling nozzle for the engine. On selecting reverse thrust, the doors rotate into the gas stream redirecting it forward. Figure 18.5 shows the propelling nozzle in the stowed for flight and activate for reverse thrust positions as performed during a pre take-off check.

Fig. 18.5

BLOCKER DOORS

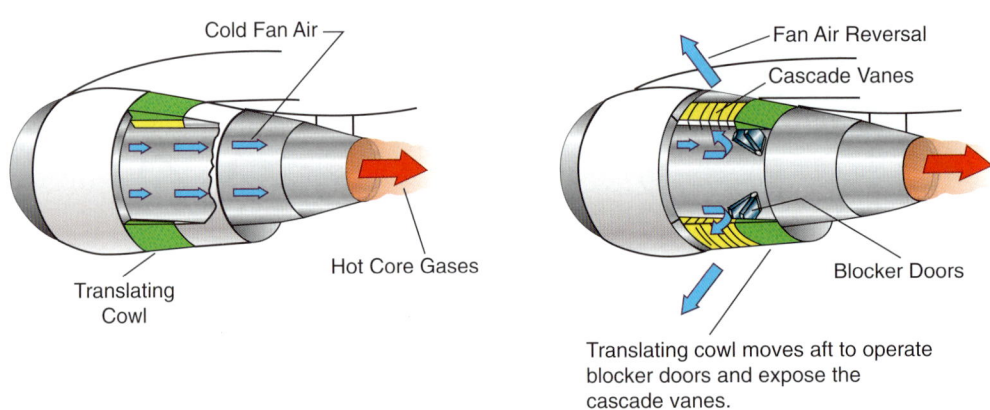

Fig. 18.6

This system is used on high bypass turbofan engines either to reverse the cold stream only or to reverse the cold and hot gas streams together. In the latter arrangement, the hot gas stream thrust reverser acts like the external door system in conjunction with the cold air stream thrust reverser.

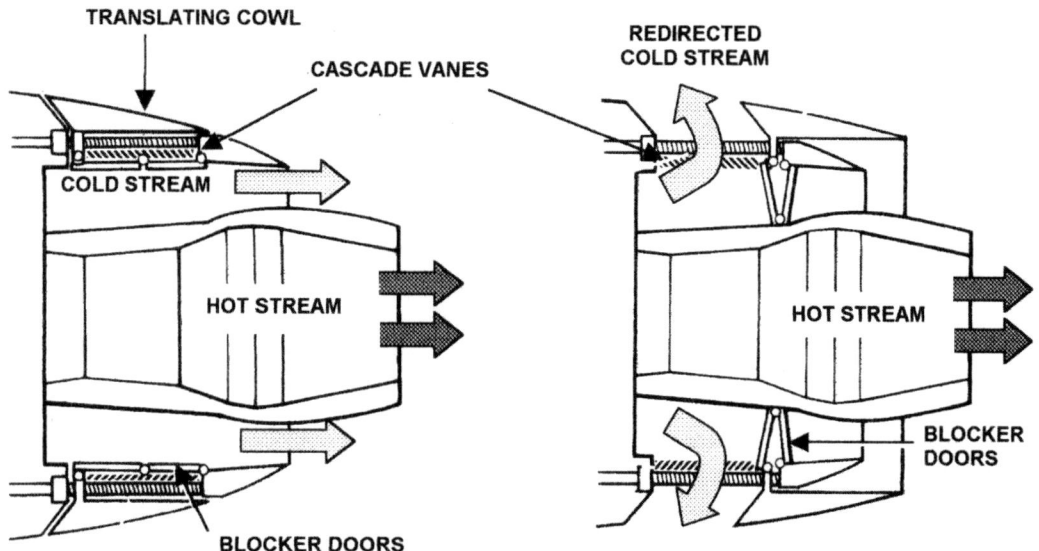

Fig. 18.7

The cold airstream thrust reverser can operate hydraulically or mechanically via an air motor. The reverser consists of a translating cowl, which in normal engine operation forms the cold air stream final nozzle, and cascade vanes that are internally covered by blocker doors. Selecting reverse thrust causes the translating cowl to move rearward, uncovering the cascade vanes and positioning the blocker doors in the air stream. As a result, this redirects the cold air stream through the cascade vanes.

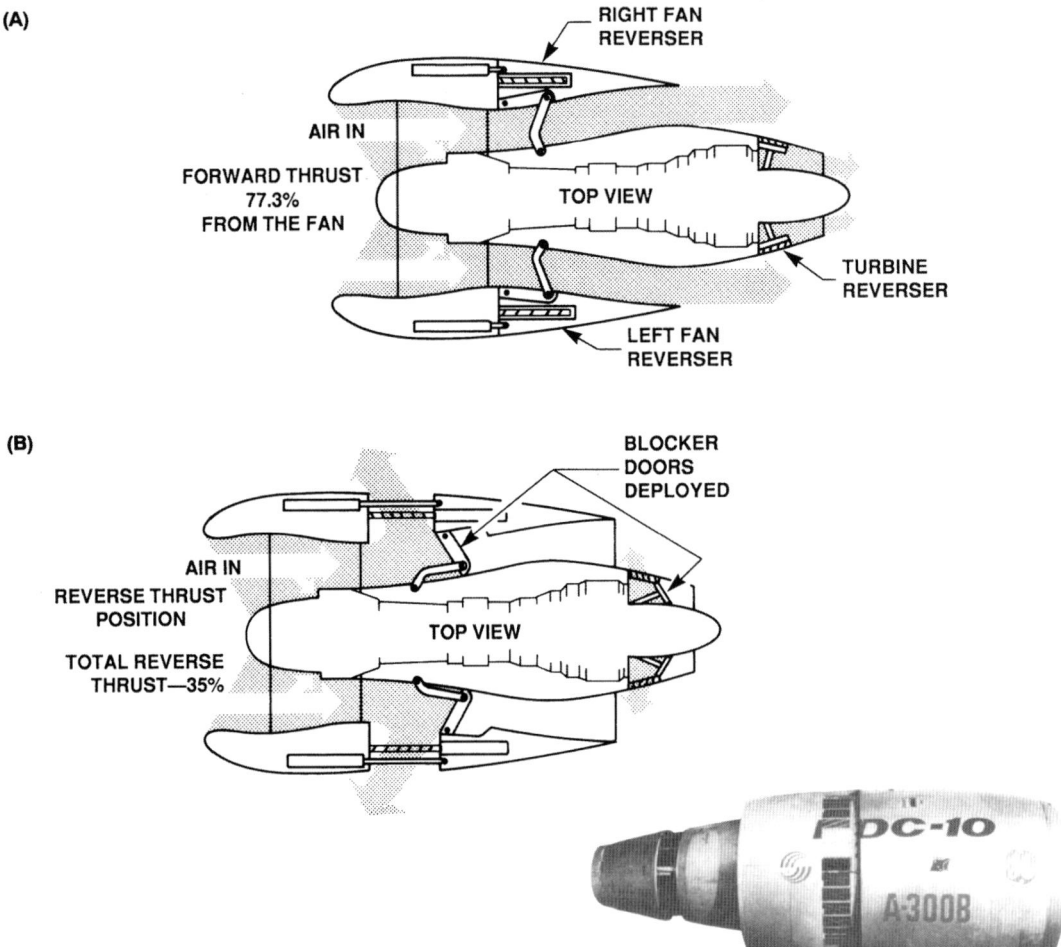

Fig. 18.8

While the cold stream is reversed, the hot stream still provides forward thrust to the aircraft. The later systems employed on modern turbofans reverse both hot and cold streams when at maximum reverse thrust setting.

OPERATION AND INDICATION

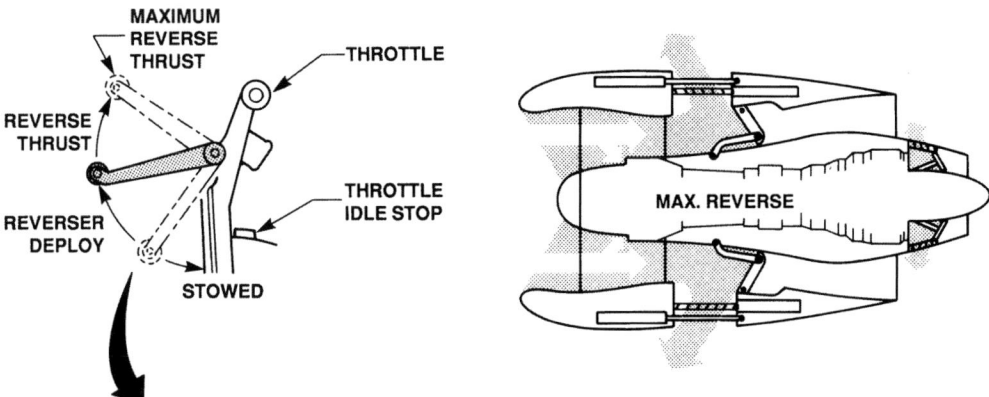

Fig. 18.9

The reverse thrust control lever is mounted on the engine thrust lever quadrant. It can either be a separate lever incorporated with the thrust lever, or be attached to its respective forward thrust lever (piggy-back lever). To operate the control lever the engines must be at idle and the aeroplane's weight must be on the wheels, which closes the ground sensing switch and allows the system to activate. Figure 18.9 illustrates a typical piggy-back lever control.

As long as the operating criteria are met, operating the control lever activates the system and deploys the reverser. The fuel flow corresponds to the lever's position, providing the correct level of reverse thrust. Depending on the system, the thrust control lever or limited movement of the normal thrust lever may control the fuel flow for reverse thrust.

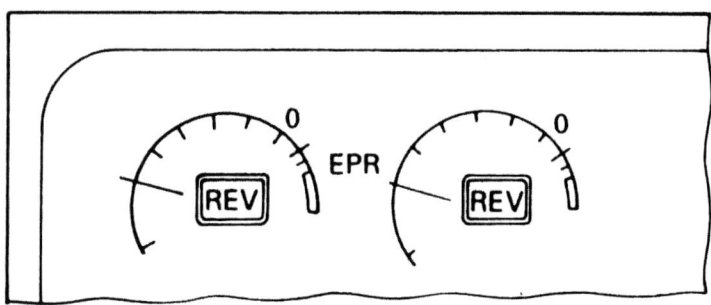

Fig. 18.10

A warning light or lights on the flight deck indicate system operation. A warning light indicates that the system is unlocked. A typical electronic display appears in figure 18.10 where the indications are as follows:

No Indication: The reverser is fully stowed and locks are fully engaged.

REV (Amber): The locks are disengaged and the reverser is between fully stowed and fully engaged (i.e. unlocked and in transit).

REV (Green): Thrust reverser is fully deployed.

Chapter 19
Gas Turbine Engine Internal Air System

INTRODUCTION

The internal air system includes those engine airflows that do not contribute directly to thrust. The system has several important functions to perform for the safe and efficient operation of the engine. These functions include internal engine and accessory unit cooling, bearing chamber sealing, prevention of hot gas ingestion into the turbine disc cavities, control of bearing axial loads, and the control of compressor and turbine tip clearances. The system also supplies air for the aircraft services.

An increasing amount of work occurs on the air as it progresses through the compressor, which raises its pressure and temperature. As a result, the air is removed as early as possible from the compressor commensurate with the requirements of each particular function in order to reduce engine performance losses. The cooling air expels overboard (via a vent system) or into the engine main gas stream, at the highest possible pressure. This achieves a small performance recovery.

COOLING

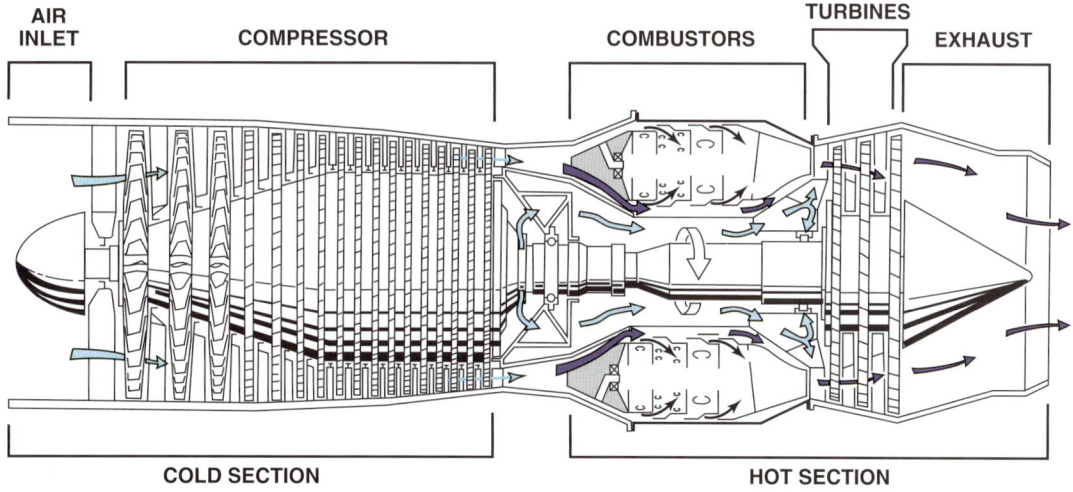

Fig. 19.1

Cooling air controls the temperature of the compressor shafts and discs by either cooling or heating them. This ensures an even temperature distribution and therefore improves engine efficiency by controlling thermal growth and thus maintaining minimum blade tip and seal clearances.

TURBINE COOLING

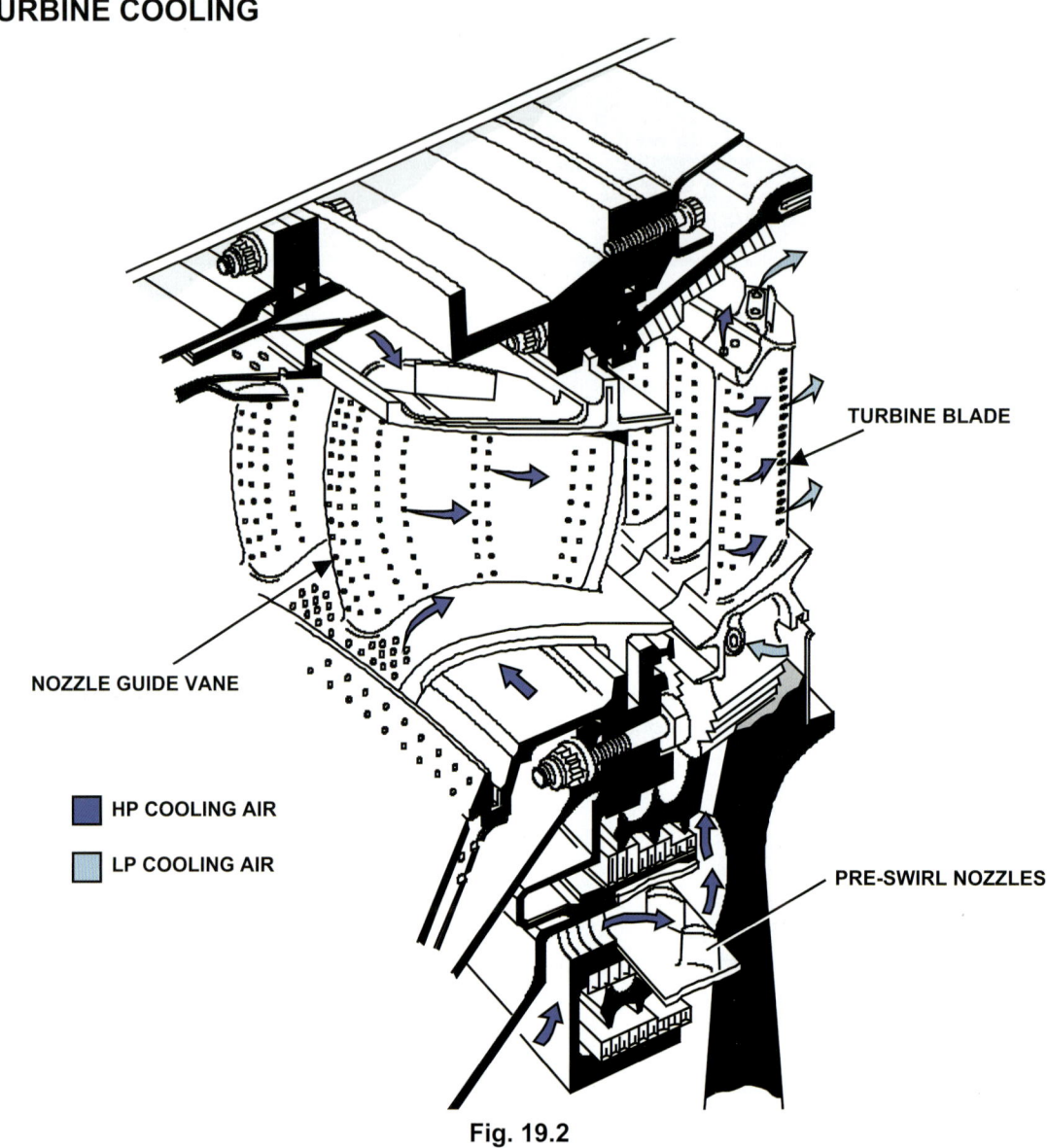

Fig. 19.2

High thermal efficiency depends on high turbine entry temperature, but these temperatures are limited by the characteristics of turbine blade and nozzle guide vane materials. Continuous cooling of the components allows their environmental operating temperature to exceed the materials' melting points without affecting the blade and vane integrity. Heat conduction from the blades to the disc requires cooling the discs, thus preventing thermal fatigue and uncontrolled expansion and contraction rates.

Cooling air for the turbine discs enters the annular spaces between the discs and flows outward over the disc faces. Interstage seals control the flow and, on completion of the cooling function, the air expels into the gas stream.

SEALING

Seals are used to prevent oil leakage from the engine bearing chambers and to control the cooling airflows, as well as to prevent ingress of the mainstream gas into the turbine disc cavities.

BEARING SEALING

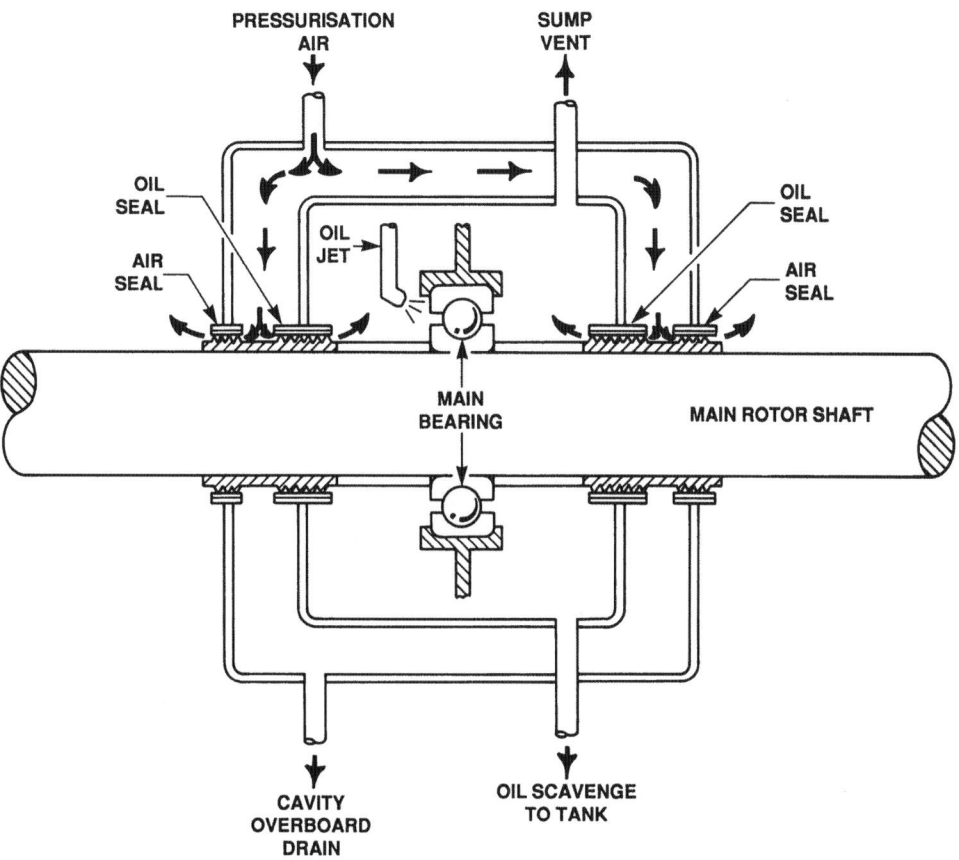

Fig. 19.3

Figure 19.3 shows a main rotor shaft bearing located within a bearing chamber and continually supplied with pressure oil via a metering jet. The bearing chamber is located within an outer chamber that is continually supplied with air under pressure. Air seals (called labyrinth seals) are located on the rotor shaft and chamber housings. These consist of a series of concentric grooves on the external surface of a ring mounted on the rotor shaft and a matching series of grooves on the inner surface of a ring attached to the chamber housing. These seals are neither airtight nor oil tight.

Since air pressure in the outer chamber is greater than the bearing chamber pressure, air flows across the faces of the air seals and into the bearing chamber, thus containing the oil. A vent dumps excess air overboard whilst scavenge pumps remove air/oil from inside the bearing chamber, after which it is returned to the oil tank.

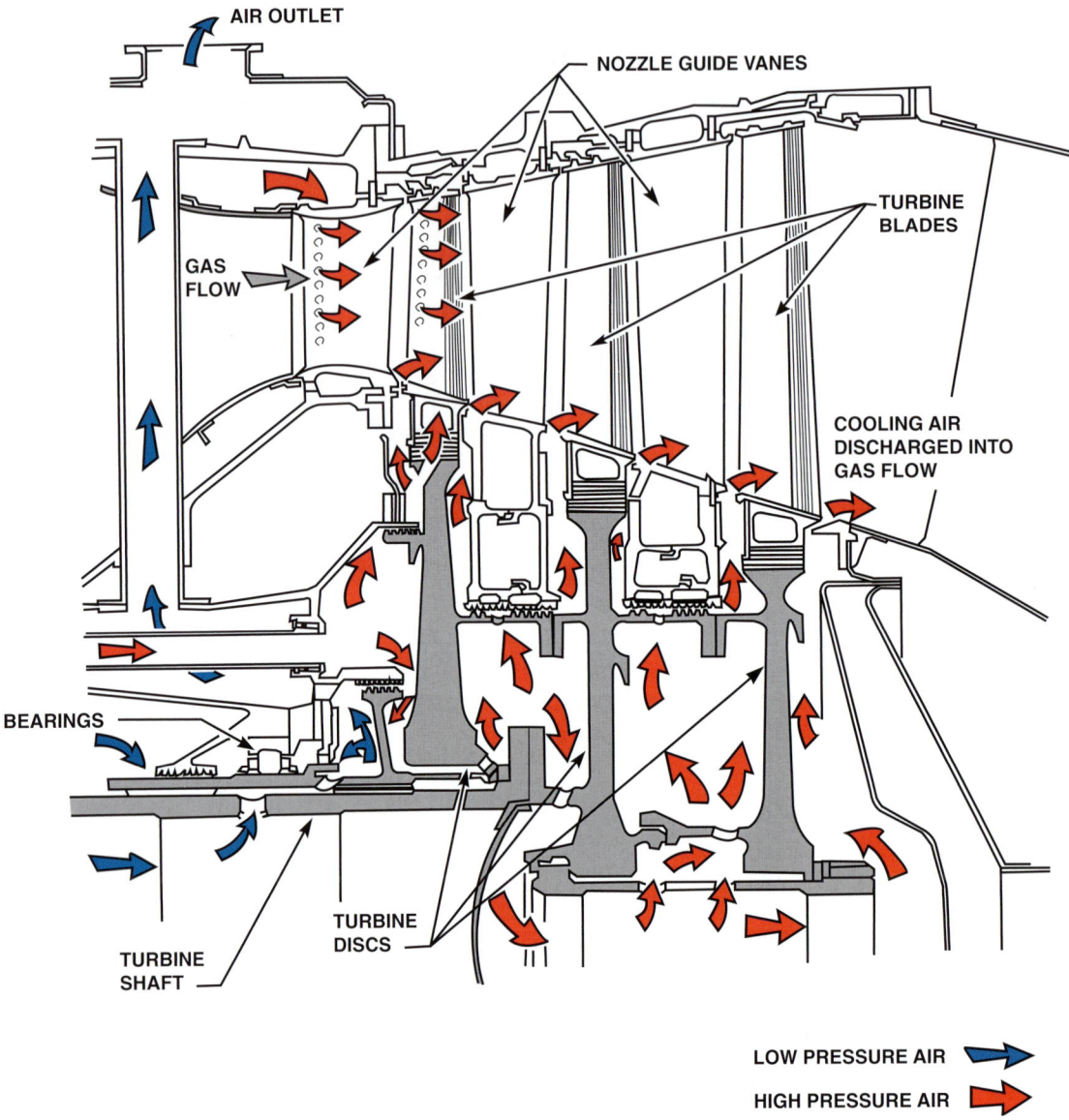

Fig. 19.4

Figure 19.4 shows how the use of shields and high-pressure air prevent gas flow from entering the core of the engine. Note the low-pressure air sealing the bearings. There are several different sealing systems for a designer to choose from. The choice of method depends mainly upon the surrounding temperature and pressure.

ACCESSORY COOLING

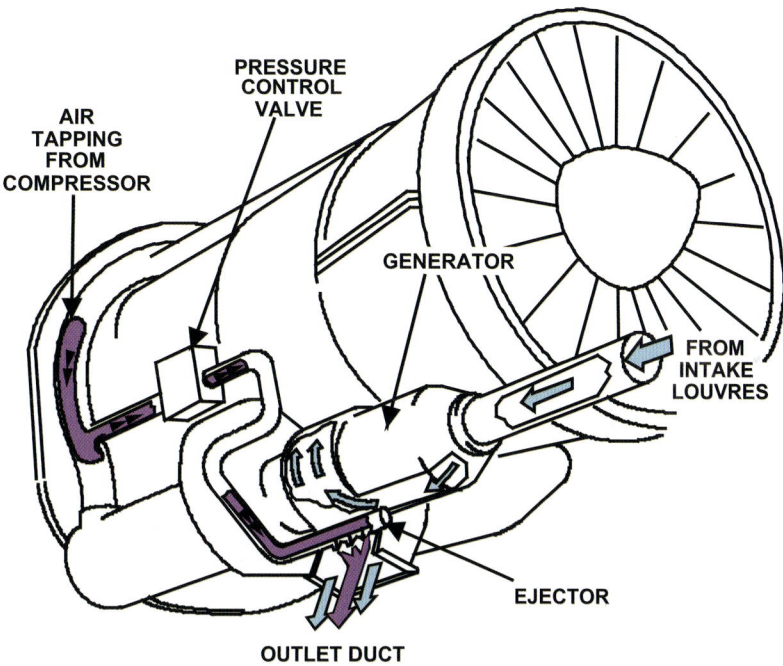

Fig. 19.5

Some of the engine accessories produce a considerable amount of heat; for example, the electrical generator. These may often require their own cooling system.

When atmospheric air-cools an accessory unit during flight, it is usually necessary to provide an induced system for use during static ground running when no external airflow would exist. Allowing compressor delivery air to pass through nozzles situated in the cooling air outlet duct of the accessory unit achieves this. The air velocity through the nozzles creates a low-pressure area that forms an ejector, thus inducing a flow of atmospheric air through the intake louvres. To ensure that the injector system only operates during ground running, a valve controls the flow of air from the compressor.

ENGINE OVERHEAT (TURBINE OVERHEAT)

On certain engines, a thermo-switch senses the temperature of the cooling air at the cooling air outlet. If the cooling air exhaust temperature reaches a specified limit, a warning appears on the flight deck. If an engine overheat occurs, the engine must be shut down immediately and not restarted.

Chapter 20
Gas Turbine Engine Gearboxes and Lubrication Systems

AUXILIARY GEARBOX

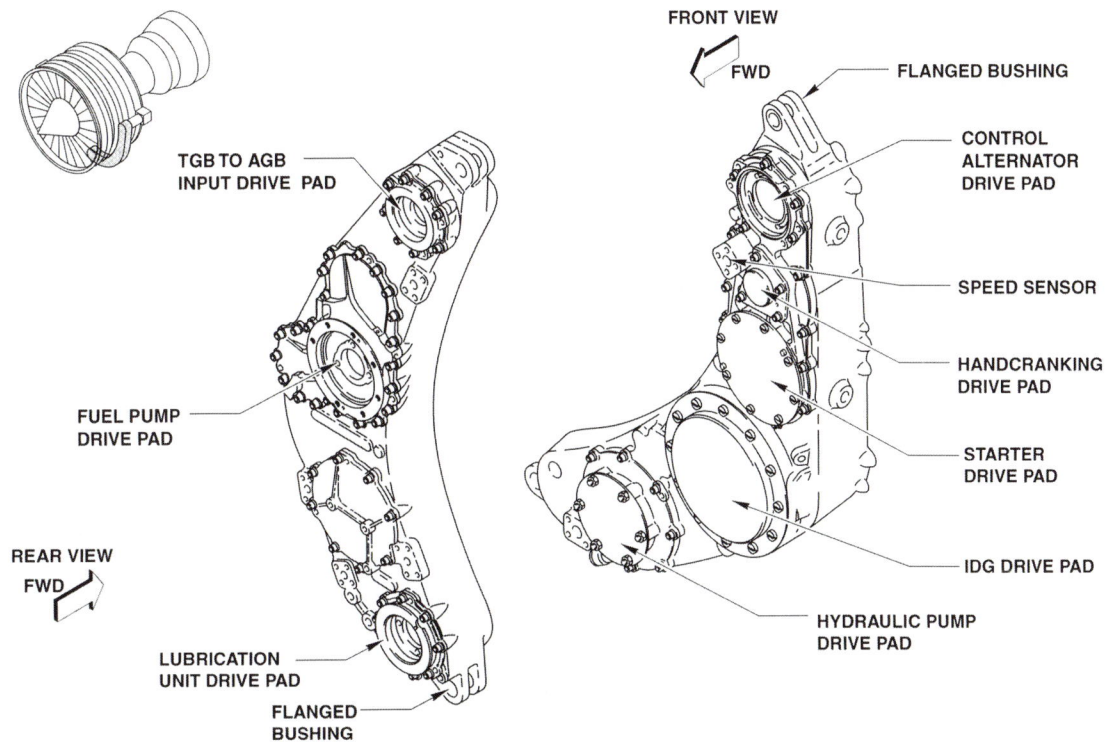

Fig. 20.1

Engines need a means of driving the accessories that power hydraulic, pneumatic, and electrical systems. Additionally, power is required for engine systems (e.g. oil pumps, fuel pumps, tacho-generators, speed governors, and dedicated alternators of a FADEC fuel control system).

GEARBOX ARRANGEMENT

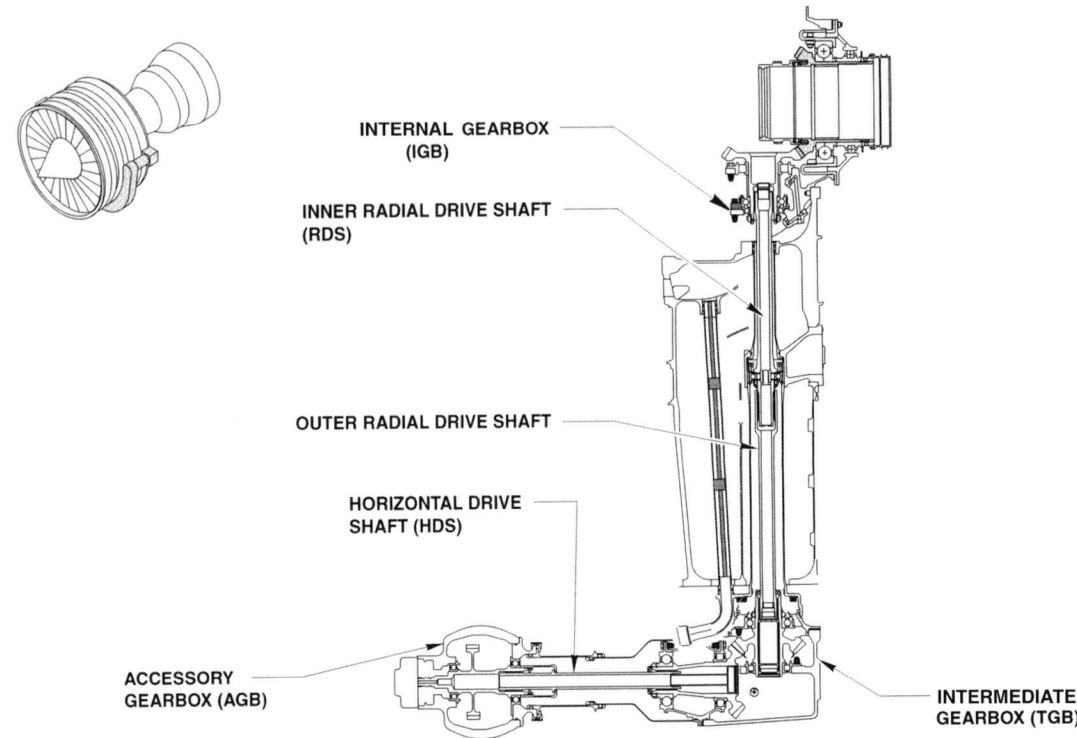

INTERNAL GEARBOX (IGB)

INNER RADIAL DRIVE SHAFT (RDS)

OUTER RADIAL DRIVE SHAFT

HORIZONTAL DRIVE SHAFT (HDS)

ACCESSORY GEARBOX (AGB)

INTERMEDIATE GEARBOX (TGB)

ACCESSORY DRIVE SECTION DESIGN

Fig. 20.2

An auxiliary gearbox is mounted externally, and normally a rotating engine shaft supplies the drive for the gearbox, where the drive transmits via an internal gearbox located within the engine core. Gearboxes are normally high speed, and on a multi-spool engine the drive comes from the HP compressor shaft. In some cases a low speed gearbox is used, and in this case the LP compressor shaft of a multi-spool engine drives it. On certain installations, a direct alignment from the rotating shaft to the gearbox may not be possible; in this case, the drive is via an intermediate gearbox that, through bevel gears, redirects the drive to the gearbox.

The lubrication requirements of gas turbine engines are generally not too difficult to meet. This is because the oil does not lubricate any parts directly heated by combustion. For satisfactory operation, an engine requires an adequate supply of oil to all bearings, gears, and driving splines. This supply must be a continuous flow of clean oil at an acceptable temperature, pressure, and viscosity, suitable for the particular application.

LUBRICATING OILS

The requirements of lubricating oil are to:

> - Lubricate
> - Cool
> - Clean
> - Prevent Corrosion
> - Resist oxidation at high temperatures
> - Possess suitable viscosity at all operating temperatures

Gas turbine engines use low viscosity synthetic oil that does not originate from mineral oil. Some early gas turbines did use a light mineral oil.

The turbojet engine is able to use low viscosity oil, due to the absence of reciprocating parts. The turbo-propeller engine requires slightly higher viscosity oil, due to the heavily loaded propeller reduction gears and the need for a high-pressure oil supply to operate pitch control mechanisms. Low viscosity oils reduce the power requirements for starting, particularly at low temperatures, with normal starts possible at –40°C.

TYPES OF SYSTEMS

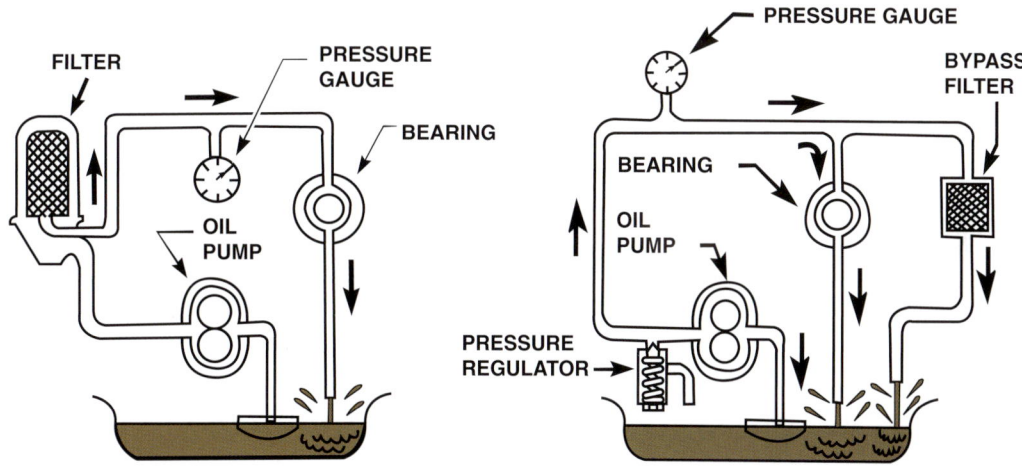

Fig. 20.3

Most gas turbines use a self-contained re-circulatory oil system of the dry sump type, where the oil distributes and returns to the oil tank via pumps. There are two basic re-circulatory systems; the **pressure relief valve** and **full flow** systems (see figure 20.3). The major difference between them is the control of the oil flow to the bearings.

The schematic diagram in figure 20.4 shows a typical full-flow system that most modern engines employ. Figure 20.5 shows the older pressure relief system.

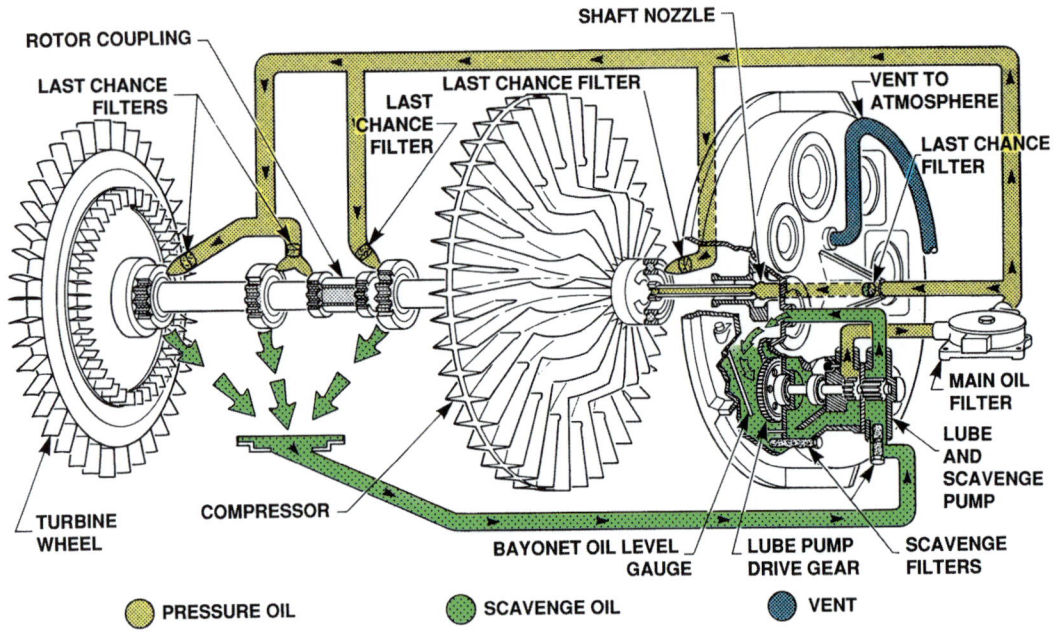

Fig. 20.4

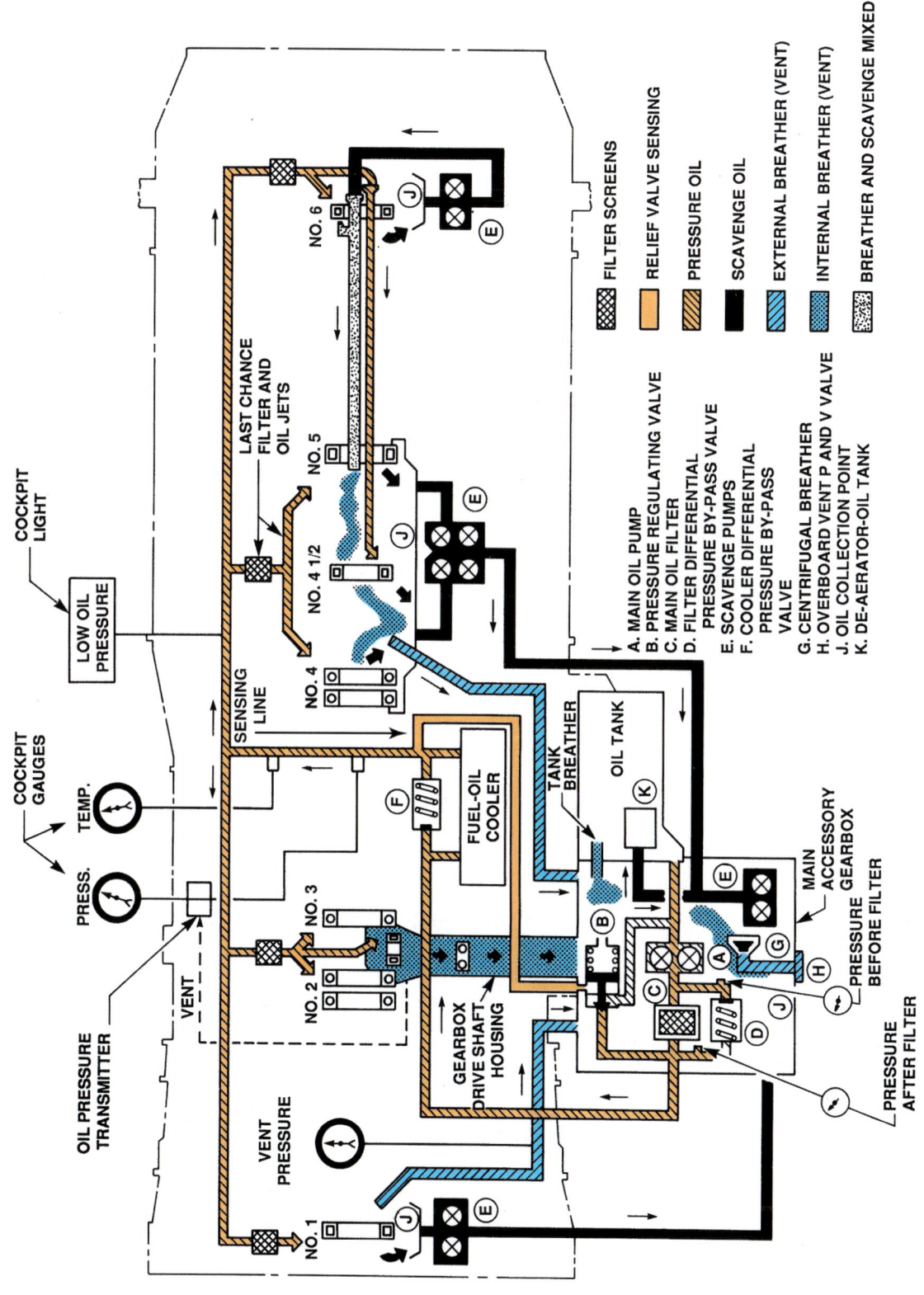

Fig. 20.5

OIL SYSTEM COMPONENTS
OIL TANK

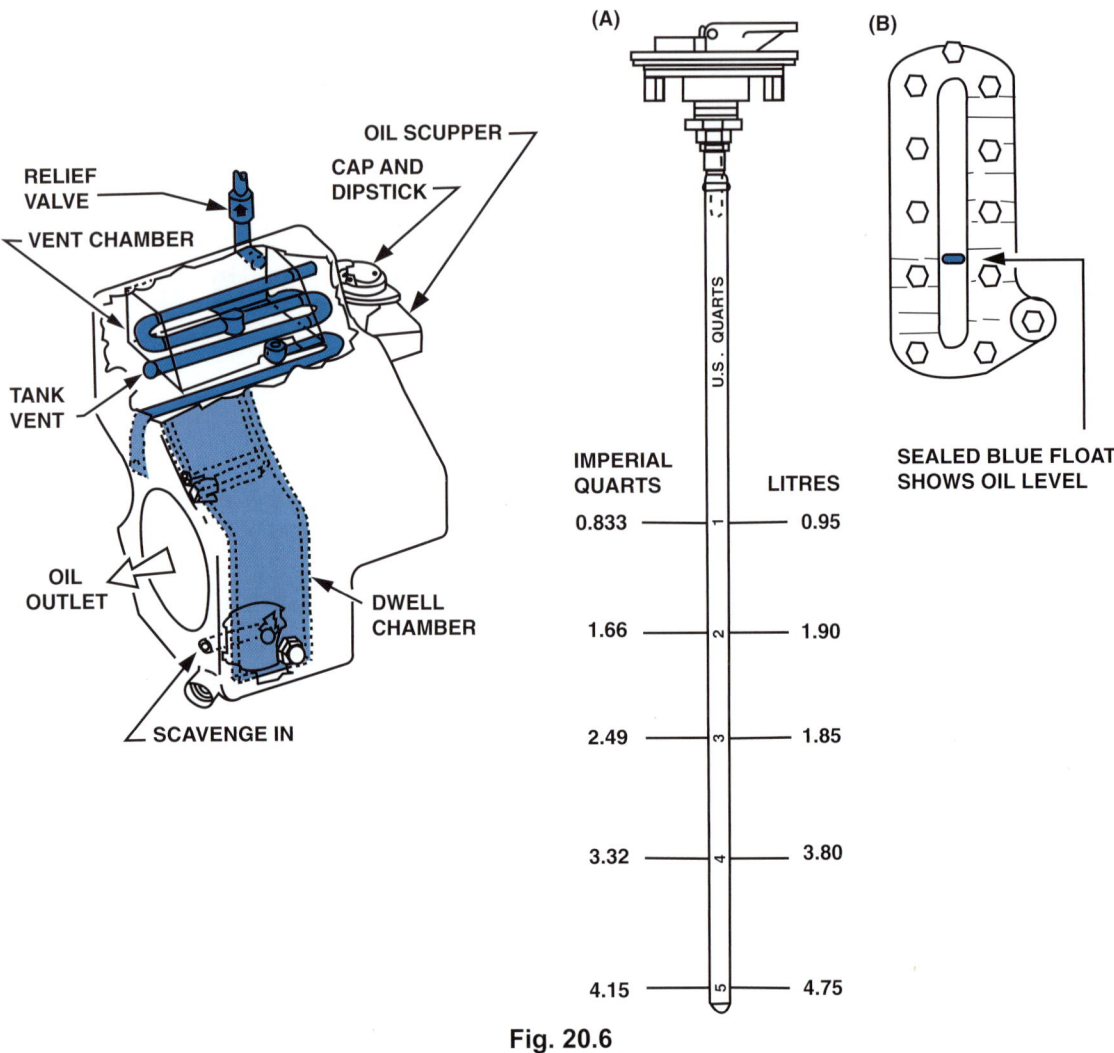

Fig. 20.6

The oil tank normally is mounted on the engine; it may be a separate unit or part of an external gearbox. It has provision for filling and draining, and has a sight glass or dipstick to allow the contents to be checked. Gravity or pressure filling replenishes the tank. To assist in removing air from the oil, the return oil passes over a de-aerator tray in the top of the tank.

FILTERS

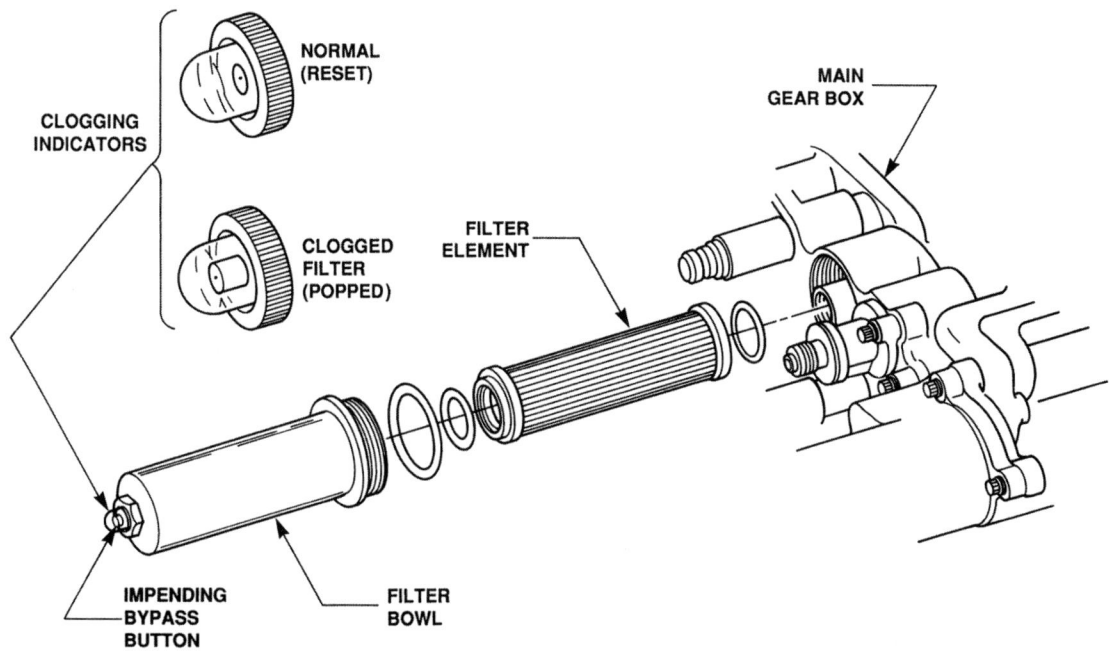

Fig. 20.7

Filters are fitted in the pressure and scavenge paths of the system. The pressure filter consists of one or more wire-wound elements to provide edge filtration and has a differential pressure switch that activates an amber filter blockage warning light on the flight deck. The scavenge strainers are normally of wire mesh construction. A fine scavenge filter attaches after the scavenge pumps and incorporates a differential pressure switch and a by-pass should the filter become blocked. Very fine thread type filters, sometimes referred to as last chance filters, are usually fitted immediately upstream of the oil jets.

OIL PUMPS

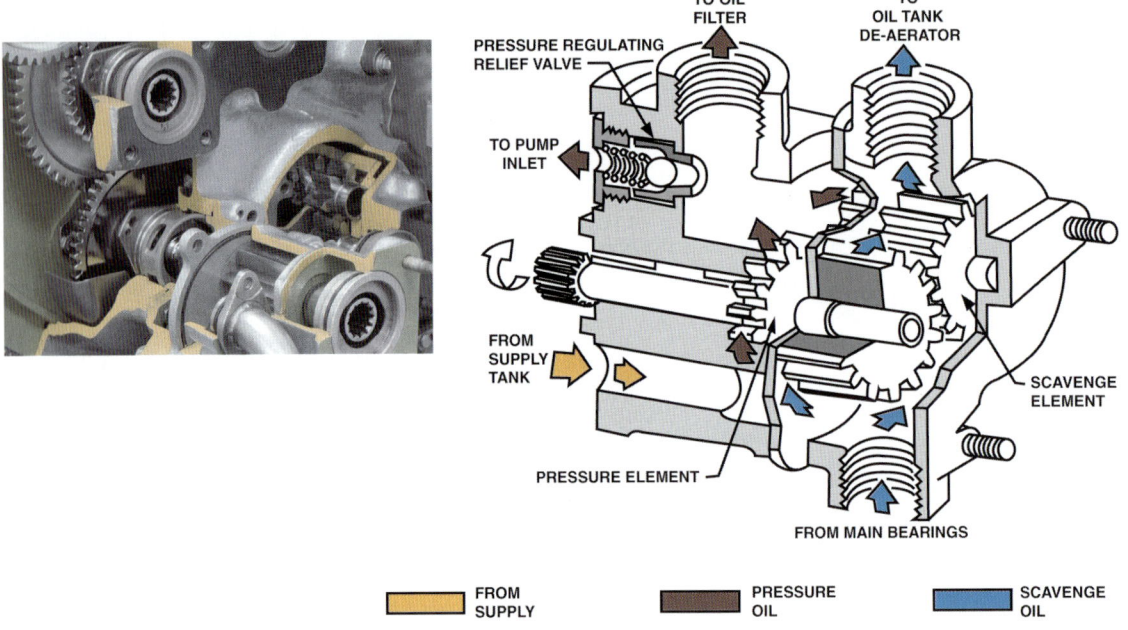

Fig. 20.8

The oil pump assembly consists of spur gear type pressure and scavenge pumps, usually fitted to a common shaft driven by the engine. The scavenge pumps have a capacity 1.5 times greater than the pressure pumps to ensure that oil is drawn from the bearing chambers and because of the expansion of the oil.

RELIEF AND BYPASS VALVES

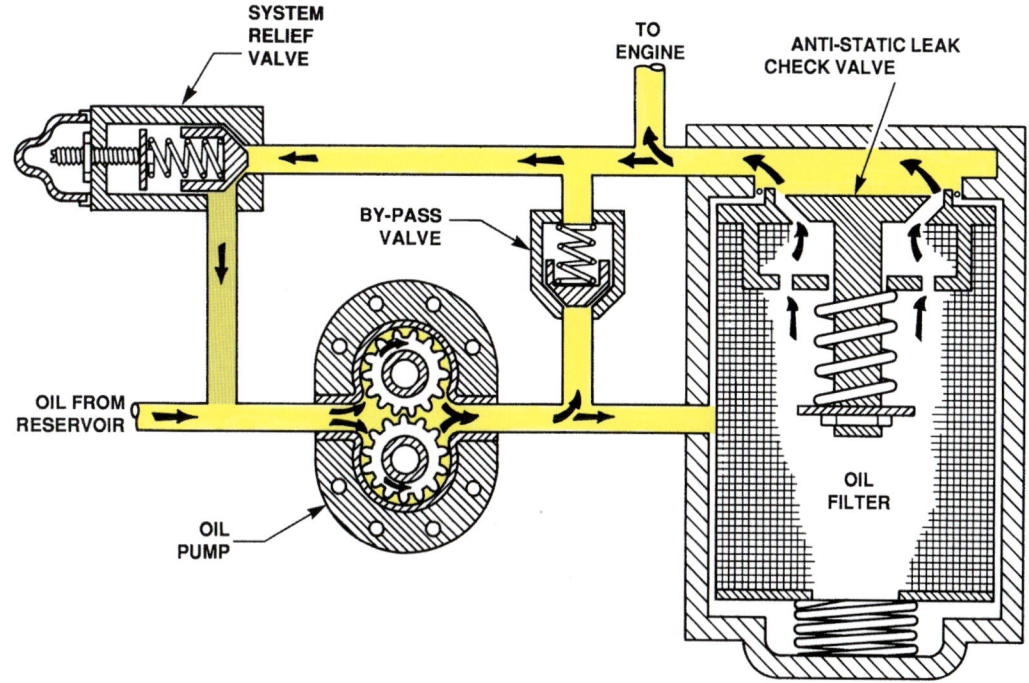

Fig. 20.9

The relief and by-pass valves are nearly always spring-loaded plate valves and do not usually contain any form of pressure adjustment.

OIL COOLERS

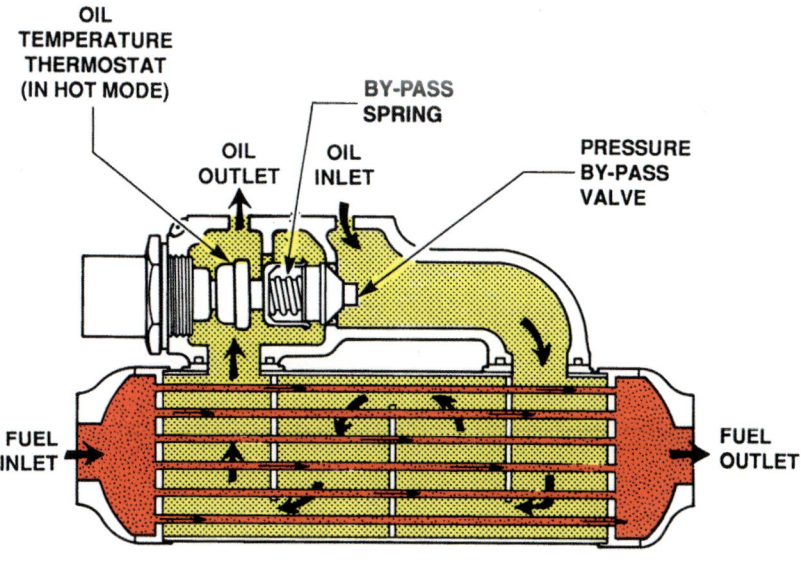

Fig. 20.10

Oil coolers consist of a matrix that is divided into sections by baffle plates; a large number of tubes convey the cooling medium through the matrix, with the oil being directed by the baffle plates in a series of passes across the tubes.

The cooling medium can be either ram air or fuel. On some air-cooled coolers, a flap valve operated automatically by the oil temperature can control the airflow through the cooler.

In some oil systems using a low-pressure, fuel-cooled oil cooler, a pressure maintaining valve is fitted. This ensures that the oil pressure through the cooler is always higher than the fuel pressure. In the event of an internal leak in the cooler, oil leaks into the fuel system rather than fuel leaking into the oil system. Some oil systems also incorporate a high-pressure fuel-cooled oil cooler. In either case, note that the fuel temperature is the controlling parameter.

CENTRIFUGAL BREATHER

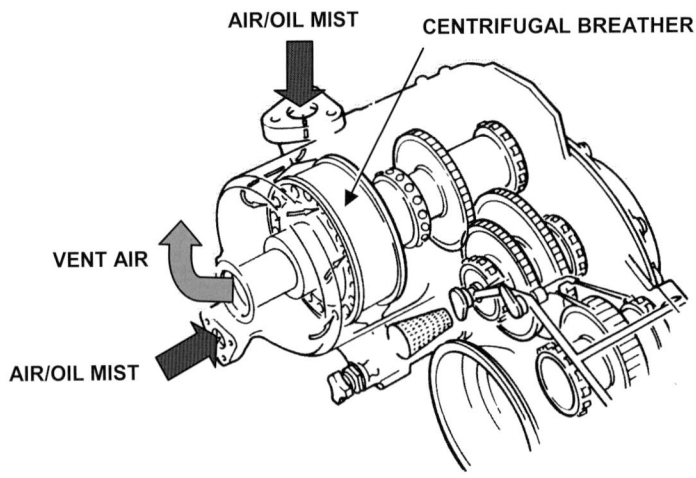

Fig. 20.11

Air is introduced into the bearing housings from the sealing system. The oil and air mixture flows over the de-aerator tray in the oil tank, where partial separation takes place. The remaining air/oil mist passes into the centrifugal breather, located on the external gearbox, for final separation. The rotating vanes of the breather centrifuge the oil from the mist, and the air vents overboard through the hollow drive shaft.

BEARINGS

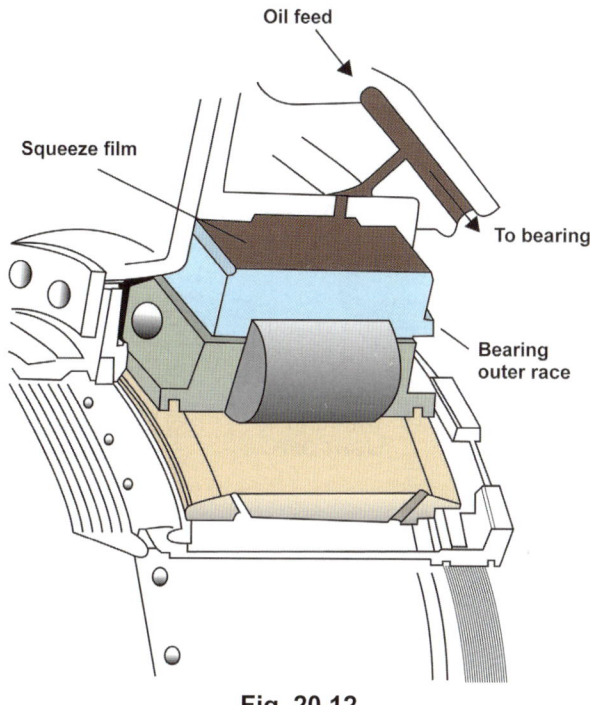

Fig. 20.12

The most common bearings used in gas turbines are the ball/roller type. To minimise the effects of the dynamic loads transmitted from the rotating assembly to the bearing housings, **squeeze film** type bearings are used.

MAGNETIC CHIP DETECTORS

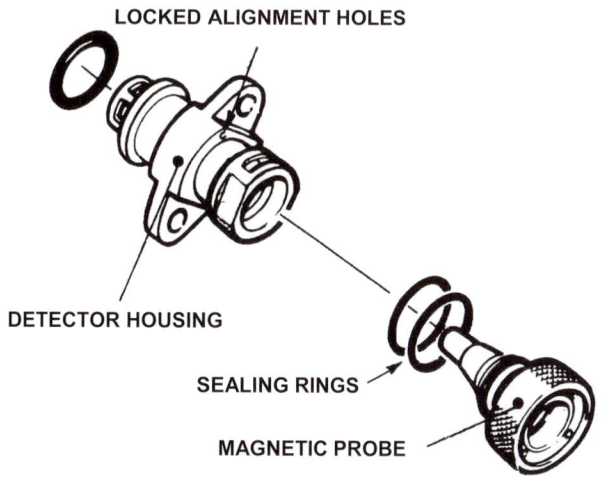

Fig. 20.13

To give early warning of bearing failure, magnetic chip detectors are fitted in the system and are located in the scavenge oil lines and gearboxes, collecting any ferrous metal particles in the oil as it returns to the oil tank. They are normally of the bayonet type fitting and can be removed, inspected, and replaced very quickly, with no oil spillage. The magnetic plugs are inspected at regular intervals for a build up of debris or indications of an impending failure.

INDICATOR CHIP DETECTOR

Normal chip detectors must be removed for inspection before any indication of wear appears; a more modern system is the indicator chip detector system. In this system, grains or chips of ferrous material (such as those given off by worn bearings) are captured from the oil stream by a permanent magnet. If a large chip or a growth of grains causes the central electrode to earth out, the flight deck receives a chip warning.

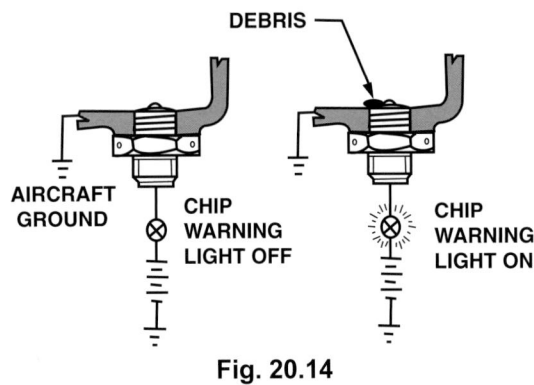

Fig. 20.14

INSTRUMENTATION

Temperature and oil pressure are critical to both systems, so these readings are indicated on the flight deck. Additional indications can include oil contents, low content warning, a red oil pressure warning light, an amber oil filter blockage warning light, and, in some cases, magnetic chip detector contamination warning.

Chapter 21
Gas Turbine Engine Fuel Systems

FUELS

Gas turbine fuel must conform to strict requirements to provide optimum engine performance, economy, safety, and overall engine life. Fuels are classified under two headings; kerosene-type fuel and wide-cut gasoline.

The commercial fuels used are:

> **Jet A-1:** This fuel has a freezing point as low as –50°C, a flash point of 38°C, and a specific gravity at 15.5°C of 0.807. This is the fuel normally used for commercial aeroplanes.

> **Jet A:** This fuel has a freezing point of –40°C, a flash point of 38°C, and a specific gravity at 15.5°C of 0.807, and is not readily available.

> **Jet B:** This wide-cut fuel has a freezing point of –60°C, a flash point of 18°C, and a specific gravity at 15.5°C of 0.764, and is not readily available.

Unlike aviation gasoline, turbine fuels are not dyed and can vary in appearance from water white to straw yellow in colour.

Low freezing points are essential due to flight at high altitudes. Most fuels contain additives to combat the problem of fuel icing. Should the temperature fall to the range where fuel icing occurs, ice crystals or a gel can form, blocking filters and components.

Fuel must also undergo a check for dissolved water, which takes on the appearance of haze or cloud in the fuel.

TYPICAL FUEL SYSTEMS

The fuel system of a gas turbine engine consists of an engine-driven pump, delivering a continuous flow of fuel to the burners in the combustion chambers, with the output of the pump varied by flow control components, to correct the flow for varying mass airflows.

Various types of fuel control systems exist fitted to a wide range of engine types that differ tremendously in their design. Regardless of the type of fuel control system employed, they all have the same basic fundamental operational requirements. A fuel system consists of two sub systems:

> Low-pressure Fuel System (LP)
> High-pressure Fuel System (HP)

Figure 21.1 shows a typical fuel system schematic:

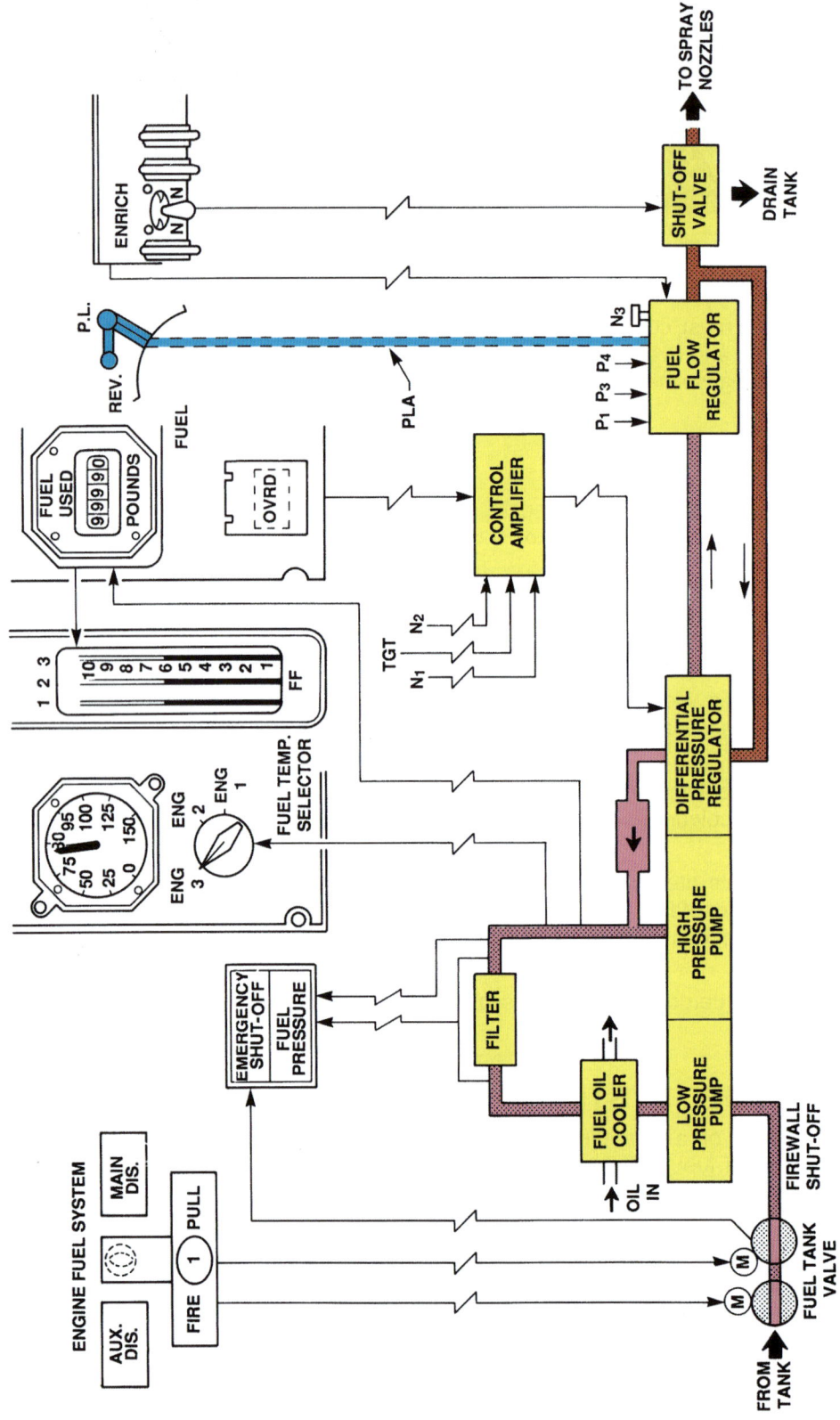

Fig. 21.1

LOW-PRESSURE FUEL SYSTEM

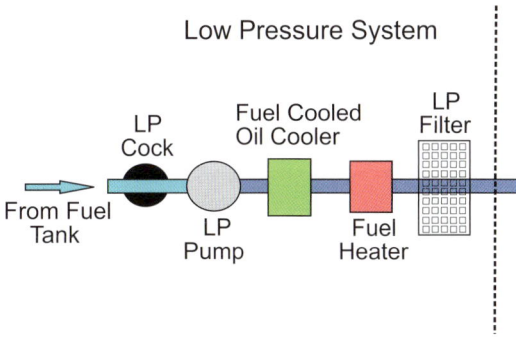

Fig. 21.2

The aeroplane tanks supply the system with fuel and condition it prior to passing it to the HP system.

LP COCK

The LP Cock, also referred to as the fuel shut-off valve (LP FSOV), is not normally closed by the flight crew other than in an emergency. This valve isolates the fuel tank (airframe) system from the engine fuel system. Should it remain closed during a start cycle, the LP and HP pumps would suffer damage as they cavitate and run dry.

LOW-PRESSURE PUMP

This pump receives fuel from the aeroplane fuel system and ensures satisfactory pressure to the high-pressure pump whilst suppressing cavitation. They are usually of the scroll type impeller that works on the induction and centrifugal principle.

FUEL HEATER

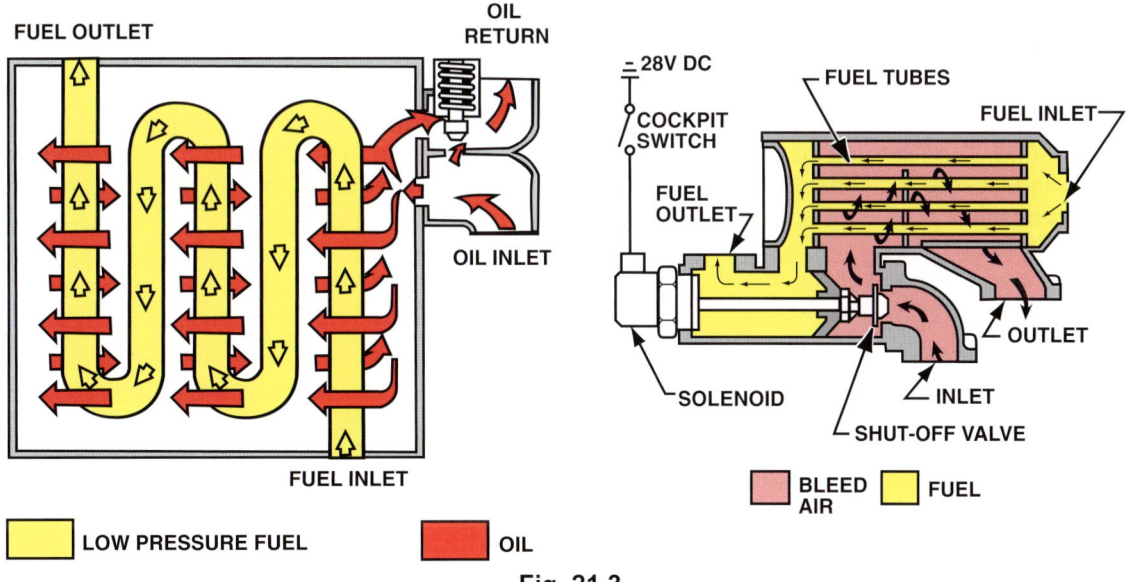

Fig. 21.3

The fuel heater is a heat exchanger that uses hot air tapped from the later stages of the compressor or hot oil from the engine lubrication system, ensuring that particles of water held in suspension in the fuel do not freeze; causing ice crystals that would lead to a filter blockage. If the heater employs hot air it is usually controlled automatically via an LP fuel temperature sensor to prevent too high a fuel temperature.

FUEL FILTER

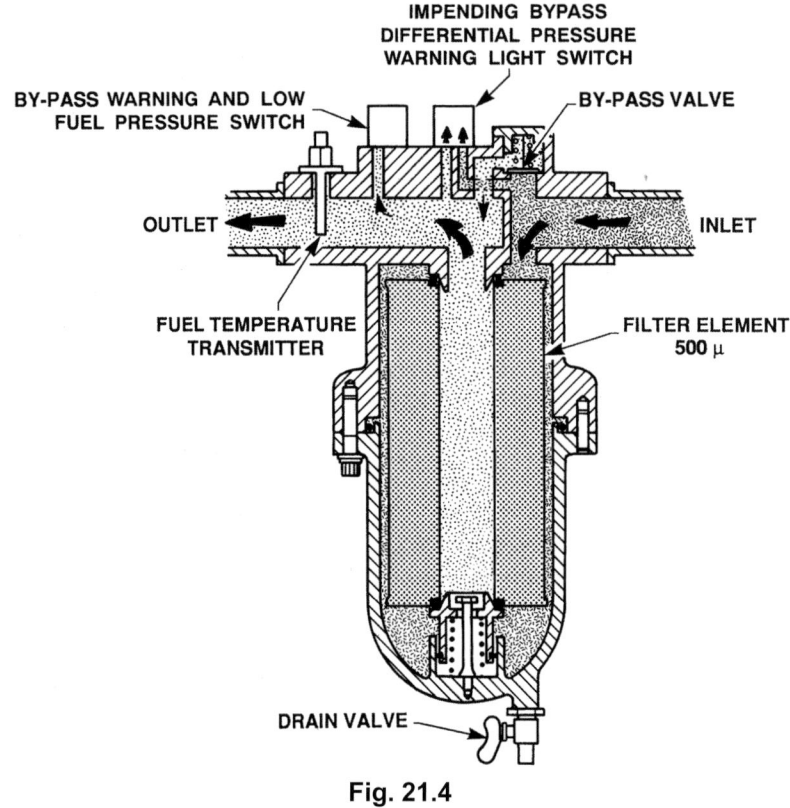

Fig. 21.4

The LP filter is normally a paper or felt element that provides fuel filtration before it passes to the HP system. The filter unit incorporates a bypass to allow the fuel to flow should the filter element become blocked. A differential pressure switch closes as the filter starts to clog, providing a flight deck indication.

HIGH-PRESSURE FUEL SYSTEM

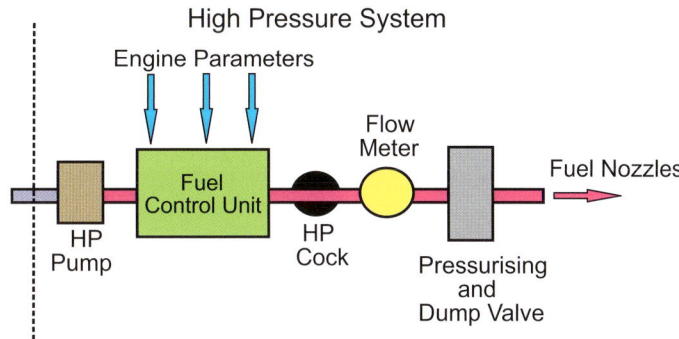

Fig. 21.5

The function of this system is to take the conditioned fuel and schedule it for use.

HIGH-PRESSURE PUMP

This pump supplies fuel at high pressure to the fuel injector nozzles to ensure good atomisation. The fuel control unit, via a spill valve, controls its output. They are either of the multi-plunger variable swash plate or gear type. Figure 21.6 shows a typical multi-plunger type pump. It has a rotor that contains several inclined cylinders containing pistons that are spring loaded against a non-rotating swash plate. A spring, acting on the servo piston, ensures that for start the swash plate is at its maximum angle, therefore ensuring maximum fuel flow.

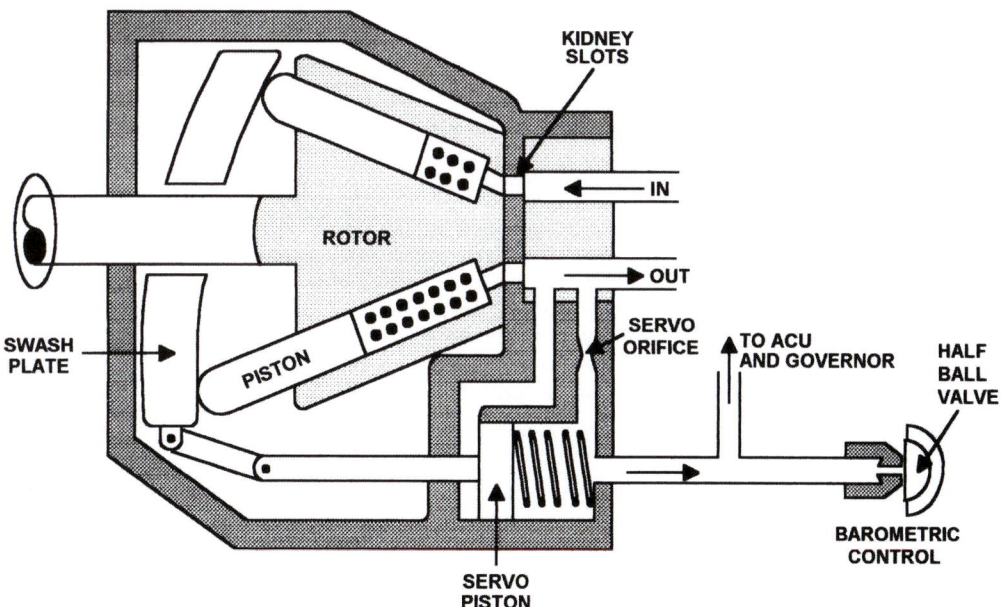

Fig. 21.6

Pump output depends on the swash plate angle and the speed of rotation. The servo piston varies the angle of the swash plate. The greater the angle, the greater the pump output and vice versa. Inlet and outlet of the pump is via kidney shaped slots in a fixed end plate. Balancing of two fuel pressures and spring pressure determines the servo position. Fuel servo and spring pressure balances the pump's output pressure. Servo pressure is obtained by pump output pressure flowing through a servo orifice, creating a pressure drop. The angle of the swash plate depends on the servo pressure, which is varied by fuel control unit internal controlling devices for such as acceleration, deceleration, airspeed, altitude, and engine speed governor.

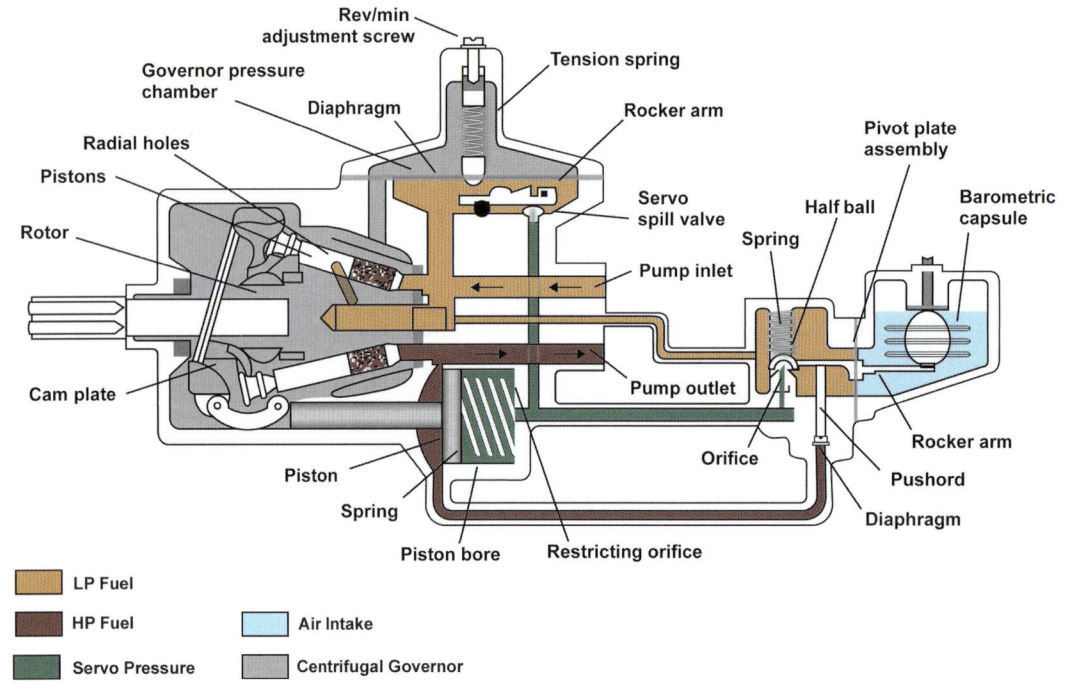

Barometric Control of Fuel Pump Output

Fig. 21.7

In figure 21.7, the barometric control varies the servo pressure for altitude changes by varying the position of a half ball spill valve. With increasing altitude, the half ball spill valve lifts from its seat, reducing servo pressure by increasing the flow through the spill valve, which in turn reduces the pump stroke and vice versa. Control for acceleration, deceleration, and engine speed governing occurs in a similar manner. Note that various types of spill valves exist, but all achieve the same aim in controlling pump output for varying conditions.

FUEL CONTROL UNIT (FCU)

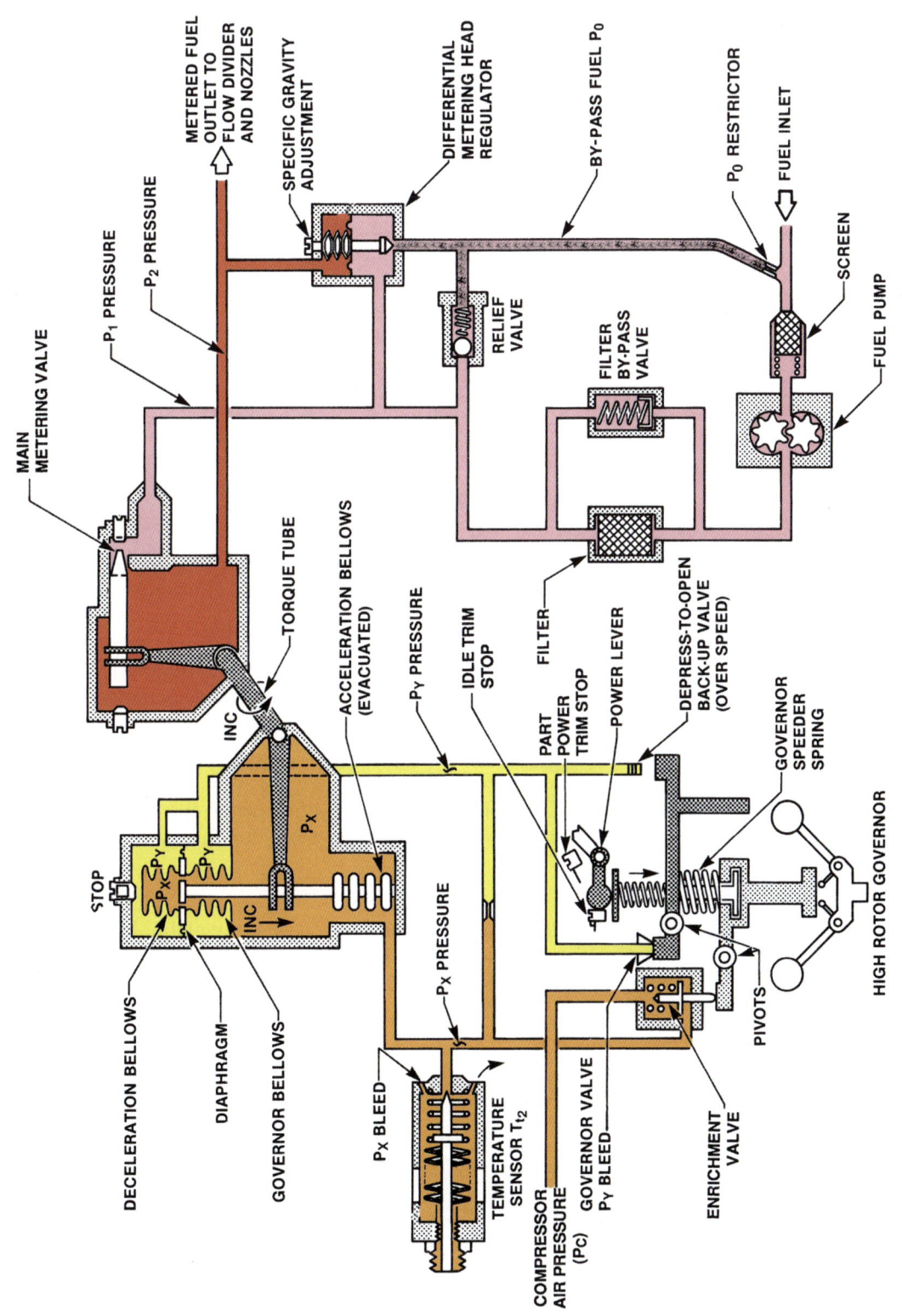

Fig. 21.8

The **fuel control unit (FCU)** controls the output of the HP pump in response to varying conditions such as acceleration, deceleration, altitude, forward speed, and engine limiting parameters. On older engines, the FCU was a mechanical device using fluid and pneumatic pressures and a governor to control the rate of fuel flow.

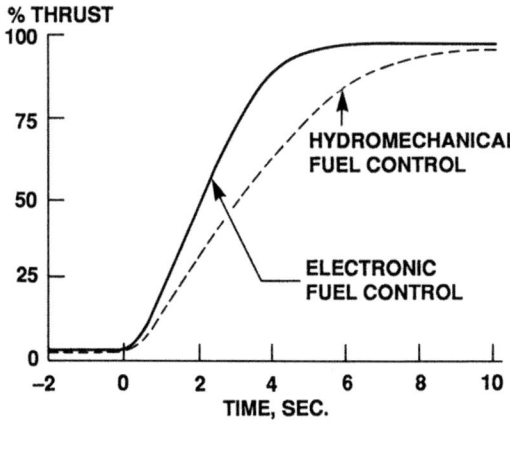

Fig. 21.9

Modern aircraft engines use electronic engine control systems to compute the fuel flow required from all the parameters, which has led to improved fuel scheduling.

HIGH-PRESSURE SHUT OFF VALVE

The **high-pressure shut off valve** is fitted between the throttle valve and the burners and serves to stop the engine by cutting off the fuel to the burners. Normally electrically actuated, this valve is usually a two-position valve (CLOSED/OPEN), but in some systems it has an intermediate (ENRICH) position which supplies extra fuel for start-up.

To prevent an excessive build up of fuel pressure in the system when the valve is closed, an idling circuit returns the HP fuel output to the LP side of the system during engine run down.

FUEL FLOWMETER

The flowmeter electrically measures the flow of fuel and indicates the weight of the fuel actually being supplied to the combustion chamber on a flight deck indicator in gallons/hour, pounds/hour, or kilograms/hour. The reading is taken from a point as close to the fuel nozzles as practicable, to give the most accurate reading.

PRESSURISATION AND DUMP VALVE

The pressurisation and dump valve ensures that sufficient fuel pressure exists before opening the high-pressure shut off valve.

FUEL INJECTOR NOZZLES

These nozzles inject the fuel into the combustion chamber in such a way as to ensure its rapid burning over the whole engine range and to provide a stable flame. They must atomise the fuel into the smallest possible droplets to obtain the maximum area for evaporation. There are two main types of burners, vaporising and atomising. Chapter 15 describes these.

FUEL CONTROL SYSTEMS

Varying the fuel flow to the combustion chamber controls the volume and velocity of the gases produced; this in turn controls engine rpm, mass airflow, and thrust. The thrust requirement is initially set by the pilot through the flight deck mounted control lever, but to achieve and maintain the required thrust without damage to the engine over a wide range of operating conditions requires a very complex engine mounted automatic fuel control system.

There are numerous types of fuel control units, some of which follow:

 - ➢ Pressure Control System
 - ➢ Proportional Flow Control System
 - ➢ Combined Acceleration and Speed Control System
 - ➢ Pressure Ratio Control System
 - ➢ FAFC Full Authority Fuel Control
 - ➢ FADEC Full Authority Digital Engine Control

The general system requirements indicate the design must allow it to operate efficiently and place minimum workload on the pilot. Thus it must respond to:

 - ➢ Movement of the throttle
 - ➢ Air pressure and temperature
 - ➢ Rapid acceleration and deceleration
 - ➢ Engine gas temperature, engine speed signals and compressor delivery pressure, ensuring that engine limitations are not exceeded

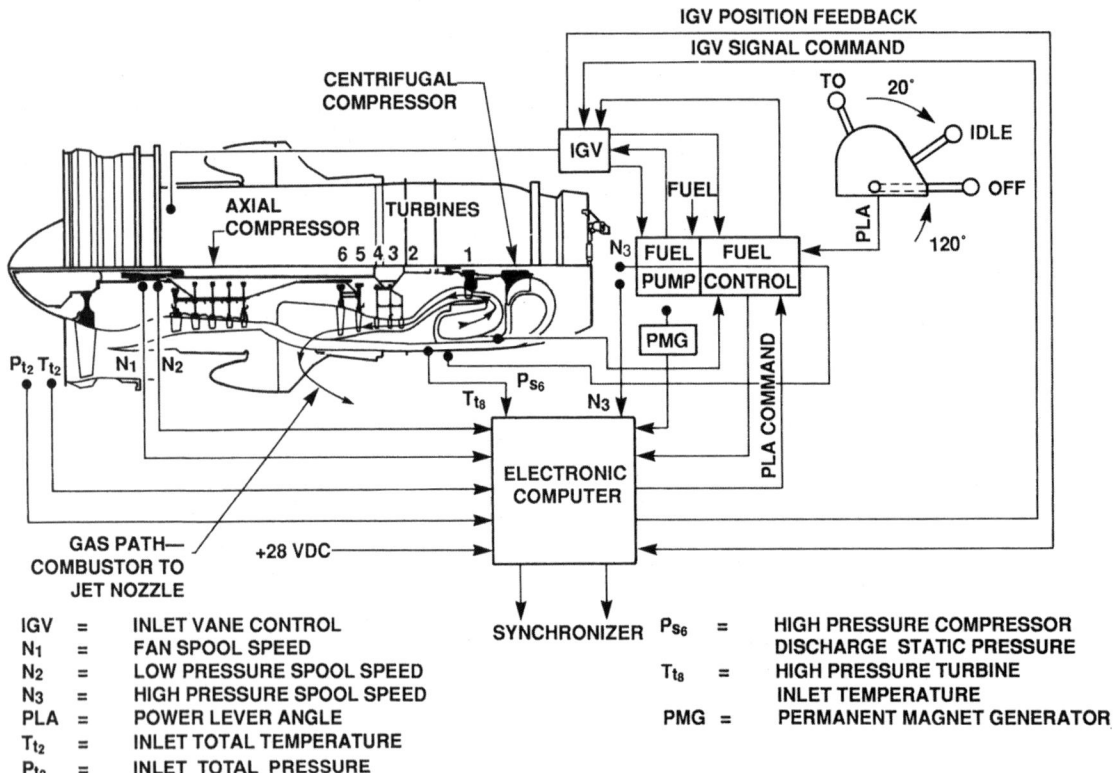

IGV = INLET VANE CONTROL
N_1 = FAN SPOOL SPEED
N_2 = LOW PRESSURE SPOOL SPEED
N_3 = HIGH PRESSURE SPOOL SPEED
PLA = POWER LEVER ANGLE
T_{t2} = INLET TOTAL TEMPERATURE
P_{t2} = INLET TOTAL PRESSURE

P_{s6} = HIGH PRESSURE COMPRESSOR DISCHARGE STATIC PRESSURE
T_{t8} = HIGH PRESSURE TURBINE INLET TEMPERATURE
PMG = PERMANENT MAGNET GENERATOR

Fig. 21.10

In the case of a FADEC system, a closed-loop system establishes engine power to control **engine pressure ratio (EPR)**, which is computed as a function of throttle lever angle, total air temperature, altitude, and Mach number. The air data computer supplies total air temperature, altitude, and Mach number to the control. Positioning the throttle aligns the control EPR command with the thrust management computer reference indicator, thus setting thrust.

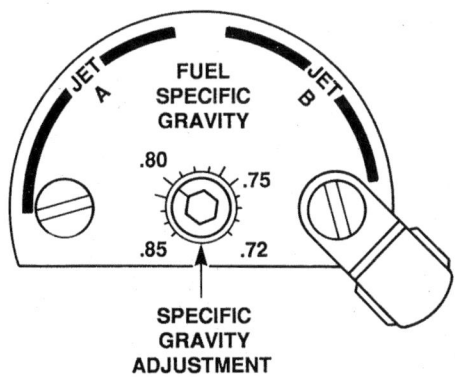

Fig. 21.11

One of the references for determining the correct fuel flow is the fuel's density; as with aviation gasoline, aviation turbine fuel has a chemically correct ratio of 15:1. As a result, should a crew uplift a different fuel from normal (e.g. Jet B instead of Jet A-1), the fuel flow rate will be incorrect since the SG is different. To compensate for this, some systems allow for an adjustment to the fuel control unit, which alters the datum, as shown in figure 21.11.

CONTROLS AND INDICATIONS

The engine fuel control system has separate controls for starting, forward thrust, and reverse thrust with indications to monitor the engine fuel system that take the form of lights and gauges. The controls and indications appear on the flight deck control quadrant and instrument panels. Some typical indications provided are dependent on the actual system, as follows:

➢ **LP Fuel Filter** — If the filter becomes clogged, a differential pressure switch illuminates a warning light.

➢ **Fuel Pressure** — A warning light illuminates when fuel pressure to the HP pump falls below a predetermined value or when the pressure drops across the filter.

➢ **Fuel Temperature** — Indicates fuel temperature at the filter outlet.

➢ **Fuel Flow** — Sensors measure the fuel flow to the engine in gallons, pounds, or kilograms per hour.

Chapter 22
Gas Turbine Engine Starting and Ignition Systems

INTRODUCTION

In order to start a gas turbine engine, it is necessary to:

➤ Rotate the engine shaft up to a speed that provides an adequate airflow. Using a starter motor achieves this.
➤ Provide fuel to mix with the airflow to provide a correct mixture for combustion.
➤ Ignite the air/fuel mixture. A high-energy ignition system provides this.

To facilitate various situations, it is necessary for the above to be controlled independently or to operate simultaneously during engine start. The starter motor can operate independently for maintenance purposes, as well as an engine dry run or ventilation run. The ignition system can operate independently for maintenance purposes and in-flight relight. Similarly, fuel control can occur independently.

TO START A TURBINE ENGINE

Starting of single-spool engines requires the whole spool to rotate. For multi-spool engines, the starter drives the HP spool (the core engine). The IP (if equipped) and LP spools start to rotate as the core engine progresses through the starting sequence.

Before it is safe to ignite fuel in the combustion chamber, the engine main shaft must be revolving at a speed that enables the compressor to provide an adequate airflow through the engine. The airflow must be sufficient to mix with the atomised fuel at the nozzles and to sustain combustion; it must also provide adequate cooling to protect the engine during **light up**.

Some form of starter motor is necessary to rotate the engine shaft up to the required self-sustaining speed. Self-sustaining speed occurs when the engine is able to support itself and accelerate without assistance from the starter motor and ignition systems. This occurs at approximately 30% N_2. The type of starter motor used varies depending on engine and aeroplane requirements. Large commercial aeroplanes normally use an air starter, whilst smaller aeroplanes usually employ an electrical starter.

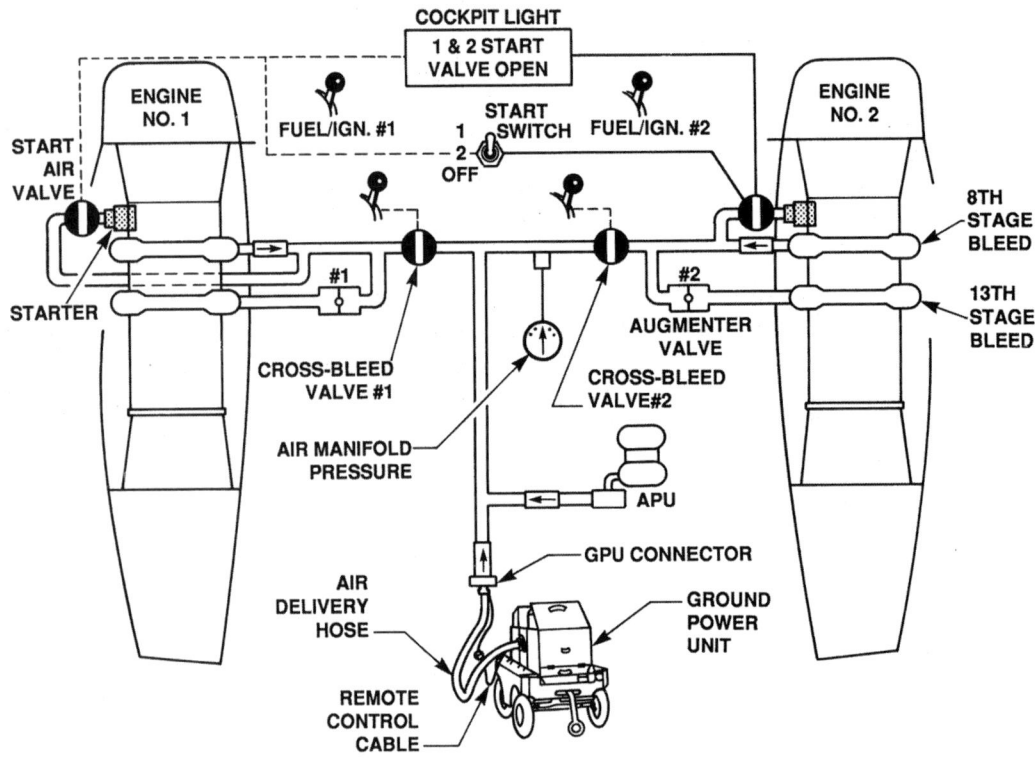

Fig. 22.1

Large commercial aircraft can use air supplied by any of the following systems as the source for the air starter (see figure 22.1):

 ➢ External ground supply
 ➢ An auxiliary power unit
 ➢ Cross-feed from a running engine

If required, a high-pressure air supply can be attached and either engine started by opening the appropriate cross-bleed valve. The air start valve opens when the pilot initiates the start sequence.

To hire a ground power unit for a start costs an operator. Most large aircraft have an internal **auxiliary power unit (APU)** (a small gas turbine engine startable by the aircraft's internal batteries). Using the APU allows the aircraft to start independently.

Having started one engine, the second engine can be started using bleed air taken from the compressor. This is not normally desirable, as the engine must be spooled up to 80% N2, which increases noise level on the stand and has the inherent danger of jet blast and suction of foreign objects into the intake. If the air manifold pressure is low due to high ambient temperatures or altitude, then extra bleed air can come from later stages of compression via the augmenter valves.

AIR STARTER

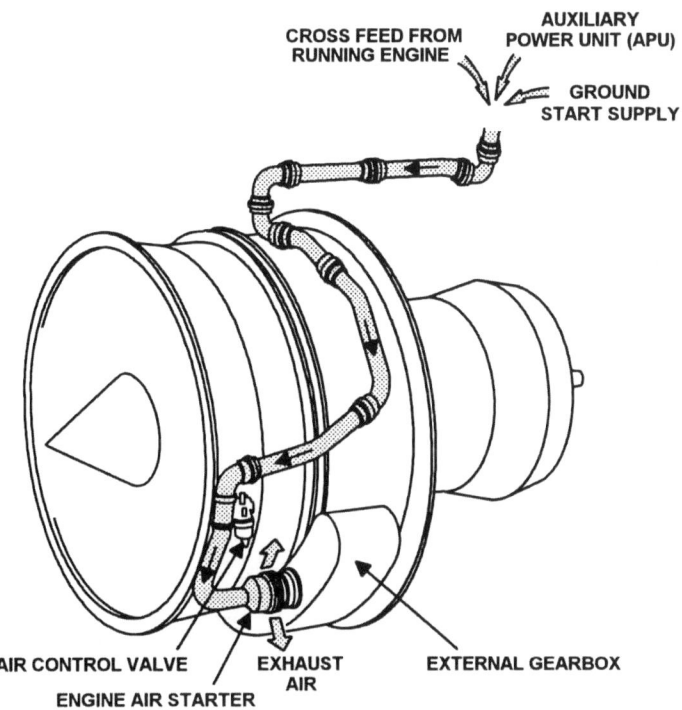

Fig. 22.2

This type of starter is a pneumatically driven turbine unit that accelerates the HP compressor shaft to the required speed for engine starting. It consists of a single-stage turbine, a reduction gear train, a clutch, and an output drive shaft, all housed within a case incorporating an air inlet and exhaust. This mounts on the engine's external gearbox and rotates the engine via the normal gearbox drive shaft.

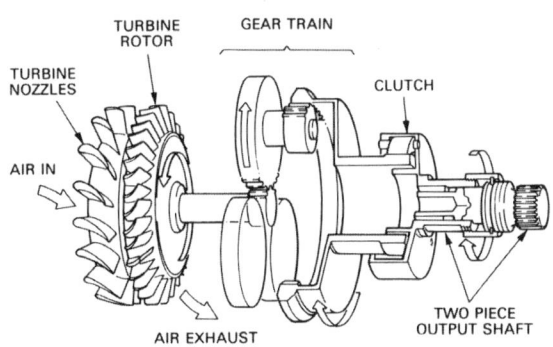

Fig. 22.3

Compressed air enters the starter, impinges on the turbine blades rotating the turbine, and exits through the exhaust. The reduction train converts the high-speed, low-torque rotation of the turbine to low-speed, high-torque rotation. The clutch engages and drives the output drive shaft, which accelerates the HP compressor shaft. When the air supply to the starter is off, the engine overruns the starter, the clutch disengages, and the starter comes to rest.

ELECTRIC STARTER

Electric starters are usually a direct current (DC) electric motor coupled to the engine via a reduction gear and clutch mechanism. After the engine has reached a self-sustaining speed, the clutch automatically disengages.

STARTING CONTROLS

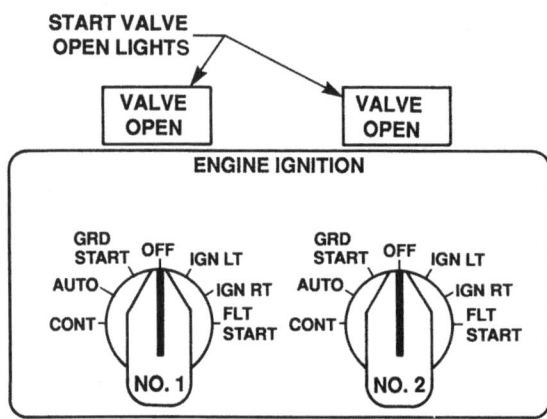

Fig. 22.4

Various control switches and indications appear on the start control panel. Depending on type, they may include such controls and indications as:

> ➢ Start Lever/Engine Master Switch
> ➢ Start Switch/Engine Start Selector/Ignition Selector
> ➢ Pneumatic Valve Operation Indicator Light
> ➢ Fire/Fault Indicator Lights

Descriptions regarding the operation of some of the controls and indications are in the typical starting cycle which follows.

TYPICAL TWIN-SPOOL TURBOFAN STARTING CYCLE

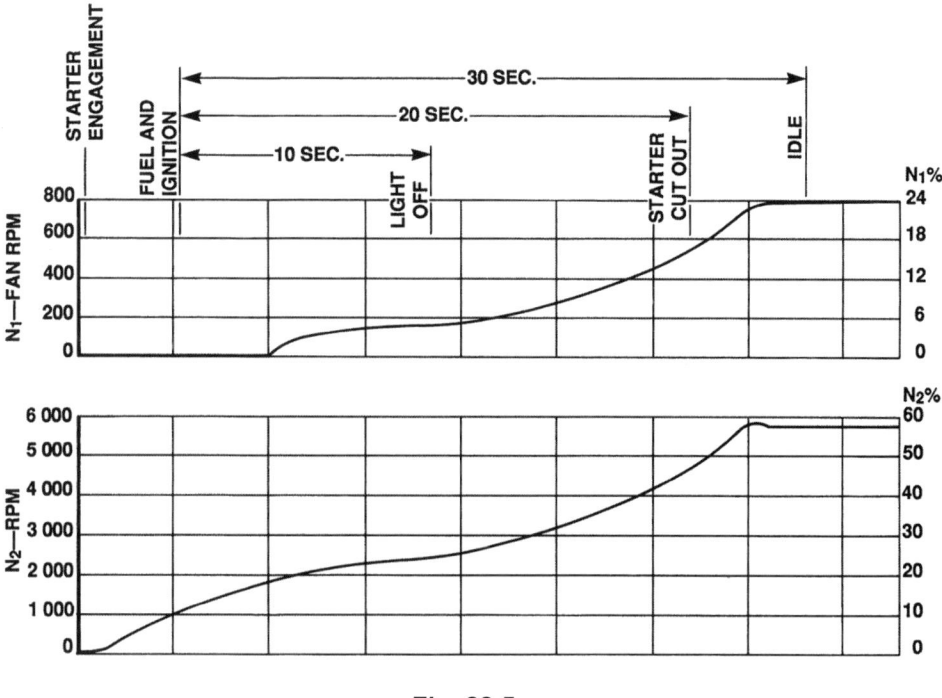

Fig. 22.5

The following is a typical staring cycle for a twin spool turbofan:

➤ **Engine master switch to start** arms the starting system.

➤ **Starter switch engagement** causes the pneumatically actuated start valve to open, indicated by a light, allowing air to flow to the starter which rotates the engine.

➤ **Ignition to ON** energises the igniters.

➤ **HP fuel cock** is placed in the **ON** position at the predetermined indicated N_2. The time to light up is observed, signified by a rise in exhaust gas temperature. The time is normally between 17 - 20 seconds. The engine accelerates to self-sustaining speed at approximately 35% N_2. The starter and ignition de-energise at approximately 45%. The engine starter switch is selected to **OFF**.

➤ **Engine acceleration** should continue up to idle once self-sustaining speed has been reached. During acceleration to idle, all engine instrumentation must be monitored to determine that no stipulated limits are being exceeded and that a satisfactory start has been achieved.

➤ **Idle Speeds** are approximately 60% N_2 and 25% N_1.

SINGLE SPOOL START CYCLE

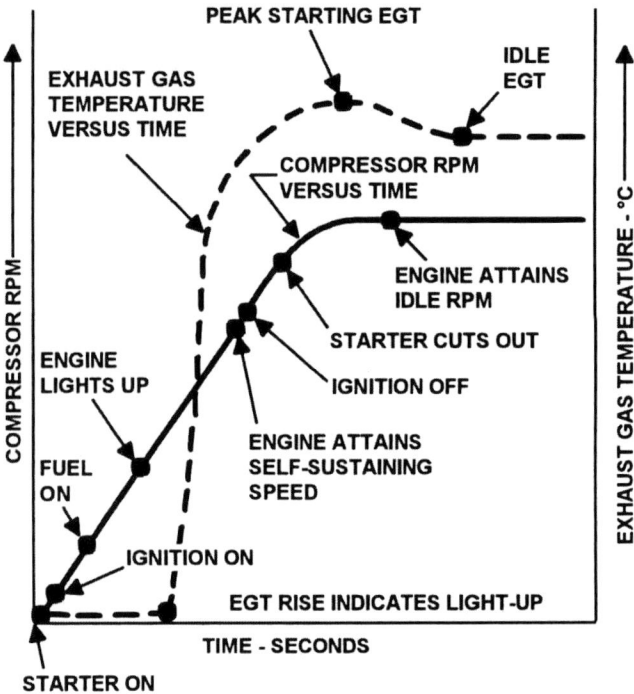

Fig. 22.6

Figure 22.6 shows that the EGT rises significantly after light up due to the combustion of the rich starting mixture from the primary nozzles and the relatively low weight of air passing through the engine. If the peak EGT exceeds the allowed maximum for the engine, turbine damage can occur.

Note that most modern starting systems incorporate automatic ignition, whereby the fuel and ignition are automatically selected during the starting cycle.

IGNITION

Fig. 22.7

Two independent high-energy ignition systems are provided, each system comprising:

  ➢ High-Energy Ignition Unit (HEIU)
  ➢ Igniter Plug (see figure 22.7 for location of an igniter plug)

Low voltage is supplied to each **high-energy ignition unit (HEIU)** and is controlled by the aeroplane's starting system electrical circuit. At a predetermined value, the stored electrical energy dissipates as a high-voltage, high-amperage discharge across the igniter plug.

Ignition units are rated in joules and provide outputs that may vary according to requirements. A high-value output of approximately 8 to 10 joules serves for satisfactory relight at altitude and starting. A low-value output of approximately 4 to 6 joules serves for continuous ignition for automatic relight should flame extinction occur due to certain flight conditions, such as icing, or take-off or landing in heavy rain, slush, or snow.

IGNITER PLUGS

Fig. 22.8

The igniter plug comes in two forms, the air gap and the surface discharge. The air gap type is similar to the conventional piston-engine spark plug, but due to the lower operating pressures in the combustion chambers has a larger air gap between the electrode and body for the spark to cross. Modern gas turbine engines normally use surface discharge plugs, as illustrated in figure 22.8. These do not have an air gap; the central electrode terminates flush with the outer case of the plug and is separated from the plug body by a semi-conductive material. When the HEIU discharges energy to the igniter plug, a flash over occurs from the central electrode across the surface of the plug to the plug body. In this system, the plug's circumference provides a larger surface area for the spark to jump from the central electrode. The advantage of this is that it prevents a single spot of carbon from stopping the spark as can happen with single point air gap spark plugs as fitted in car engines. A functional check for the igniter plugs is to press the re-light switch and listen for a crack as the spark occurs.

IGNITION MODES OF OPERATION

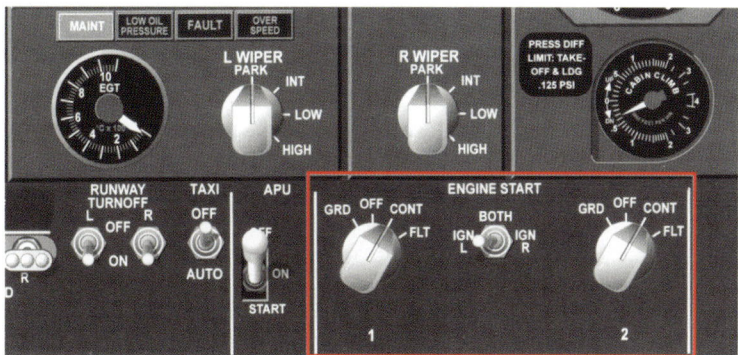

Fig. 22.9

The ignition system is required to satisfy various operating conditions, such as:

> ➢ Ground start
> ➢ In-flight start
> ➢ Continuous ignition
> ➢ Automatic ignition

GROUND START

This is the normal mode for engine start. On selecting start, the igniters operate, discharging at a rate of normally approximately 60 to 100 sparks per minute at the high-energy level of approximately 8 to 10 joules.

IN-FLIGHT START

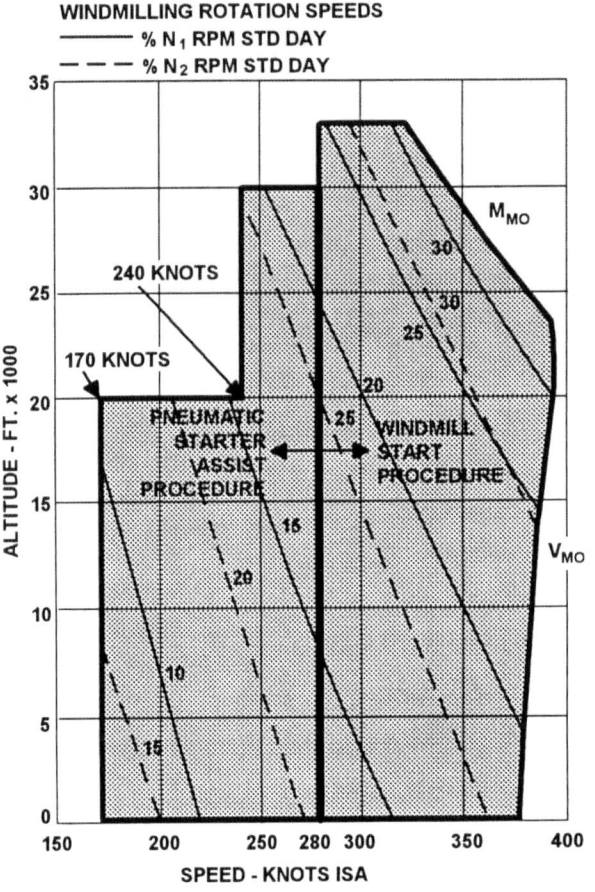

Fig. 22.10

Should the engine require relighting in flight due to combustion being extinguished, then provision for relight must be available. To ensure an in-flight relight the aeroplane may have to descend to a specified altitude and airspeed. An in-flight relight normally does not require the assistance of the starter motor as windmilling of the compressor gives the required rotation. However, under certain circumstances use of the starter motor may be required. Once the correct conditions are met with fuel available, the relight switch can be activated turning on the ignition. Figure 22.10 shows a typical in-flight relight envelope.

CONTINUOUS IGNITION

Continuous ignition is used when there is a danger of flame extinction in the event of icing, take-off or landing in heavy rain, slush, or snow and can be selected either manually or, on some installations (such as the V2500 fitted to the Airbus A320) automatically. The manual operation requires the system to be switched ON and it will remain ON until switched off. Automatic selection occurs when the engine anti-ice system is ON or when the aeroplane flaps are extended for take-off, approach, and landing. Continuous ignition operates at the low-energy level of approximately 4 to 6 joules.

AUTOMATIC IGNITION

On some installations fitted with electronic engine control, automatic ignition can serve for a normal engine start where the electronic engine control monitors the engine speed and exhaust gas temperature. If a hung or hot start is detected, the fuel, ignition, and start air automatically shut off. Automatic selection of continuous ignition as described above is the normal run position after engine start in an automatic ignition system. In other installations, the automatic ignition may link to the stall warning system of an aeroplane and activate the ignition system as the stall approaches deactivating the system as the aeroplane moves away from the stall.

ENGINE START MALFUNCTIONS

During an engine starting cycle, various start malfunctions can occur that would prevent a satisfactory start. It is essential that the engine instruments are monitored throughout the cycle in order to prevent a potentially dangerous situation developing resulting in engine or aeroplane damage. Should an unsuccessful start occur, restart attempts are usually limited to three over a specific time period with a rest period between each attempt.

WET START

Wet start is when the engine does not light up within the specified period with no indication of a rise in exhaust gas temperature, no increase in rpm, no sound indicating that the fuel being sprayed into the combustion chamber has ignited, or an abnormally low fuel flow. The causes of a wet start may be:

> ➢ Faulty high-energy ignition unit
> ➢ Faulty igniter plug
> ➢ Internal start — battery voltage low

After a wet start, it is important to dry out the engine before attempting another start. It may be necessary for the engine to be motored over by the starter motor only without fuel or ignition to remove the excess fuel in the combustion chamber, turbine, and jet pipe. Depending on the specific type of system, the motoring selection can be identified on the starter control panel as **Vent Run**, **Dry Run**, **Blow Out**, or **Motoring Run**.

HOT START

This occurs after light up and the exhaust gas temperature exceeds the maximum allowable starting temperature. The primary causes of a hot start are:

> ➢ Low electrical power supply that cannot bring the engine up to the self-sustaining speed quickly enough.
> ➢ Low air pressure to the air starter.
> ➢ Failure to allow complete draining and drying of the engine after a wet start.
> ➢ A strong tail wind into the jet pipe.
> ➢ Early opening of HP cock.

As soon as a hot start becomes apparent, the HP fuel cock should be CLOSED and the starting cycle terminated.

HUNG START

This occurs after achieving light up, but the rpm does not increase to that of idle, remaining at some lower rpm with the exhaust gas temperature at some value below or equal to the starting limit, which is high for that rpm. The possible causes of a hung start are:

> ➢ Fuel control malfunction
> ➢ Premature starter disengagement
> ➢ Shaft bearing failure

As in the case of a hot start, the HP fuel cock should be CLOSED and the starting cycle terminated.

Chapter 23
Gas Turbine Engine
Electronic Engine Control

INTRODUCTION

The evolution of gas turbine technology demanded more precise control of engine parameters than the abilities of conventional hydro-mechanical systems.

The first **electronic engine control system (EEC)** was a supervisory control. The supervisory control system combines with the proven hydro-mechanical controls. The major components in the supervisory control system include the control itself, the fuel control of the engine, and the bleed air and variable stator vane control.

With the supervisory control, the pilot simply moves the thrust lever to a desired thrust or maximum climb position. The control adjusts **engine pressure ratio (EPR)** as required to maintain the thrust rating in spite of changes in flight and ambient conditions. The control also limits engine speed and temperature, ensuring safe operation throughout the flight envelope.

If a problem occurs in this system, control automatically reverts to the hydro-mechanical system, without discontinuity in thrust. The pilot can also revert to the hydro-mechanical systems at any time.

This control led to the full authority EEC, which is fully redundant, controls all engine functions, and eliminates the need for the back-up hydro-mechanical control used in the supervisory systems. The full authority EEC is called full authority digital engine control (FADEC).

FULL AUTHORITY DIGITAL ENGINE CONTROL (FADEC)

One of the basic purposes of FADEC is to reduce flight crew workload, particularly during critical phases of flight. The FADEC's control logic achieves this, simplifying power setting for all engine operating conditions. The thrust levers achieve engine thrust values at constant lever positions, regardless of flight or ambient conditions. For example, assuming a given EPR at a particular OAT, changing the OAT causes the system to adjust the engine fuel system accordingly to maintain the EPR.

The FADEC establishes engine power through direct closed-loop control of EPR, which is the thrust rating parameter. Selection of EPR is calculated as a function of thrust lever angle, altitude, Mach number, and total air temperature.

The Air Data Computer supplies altitude, Mach number, and total air temperature to the control. Sensors provide measurements of engine temperatures, pressures, and speeds, and this data serves to provide automatic thrust rating control, engine limit protection, transient control, and engine starting. The control implements EPR schedules to obtain the EPR rating at various throttle lever angle positions and provides the correct rating at a constant throttle lever angle during changing flight or ambient conditions.

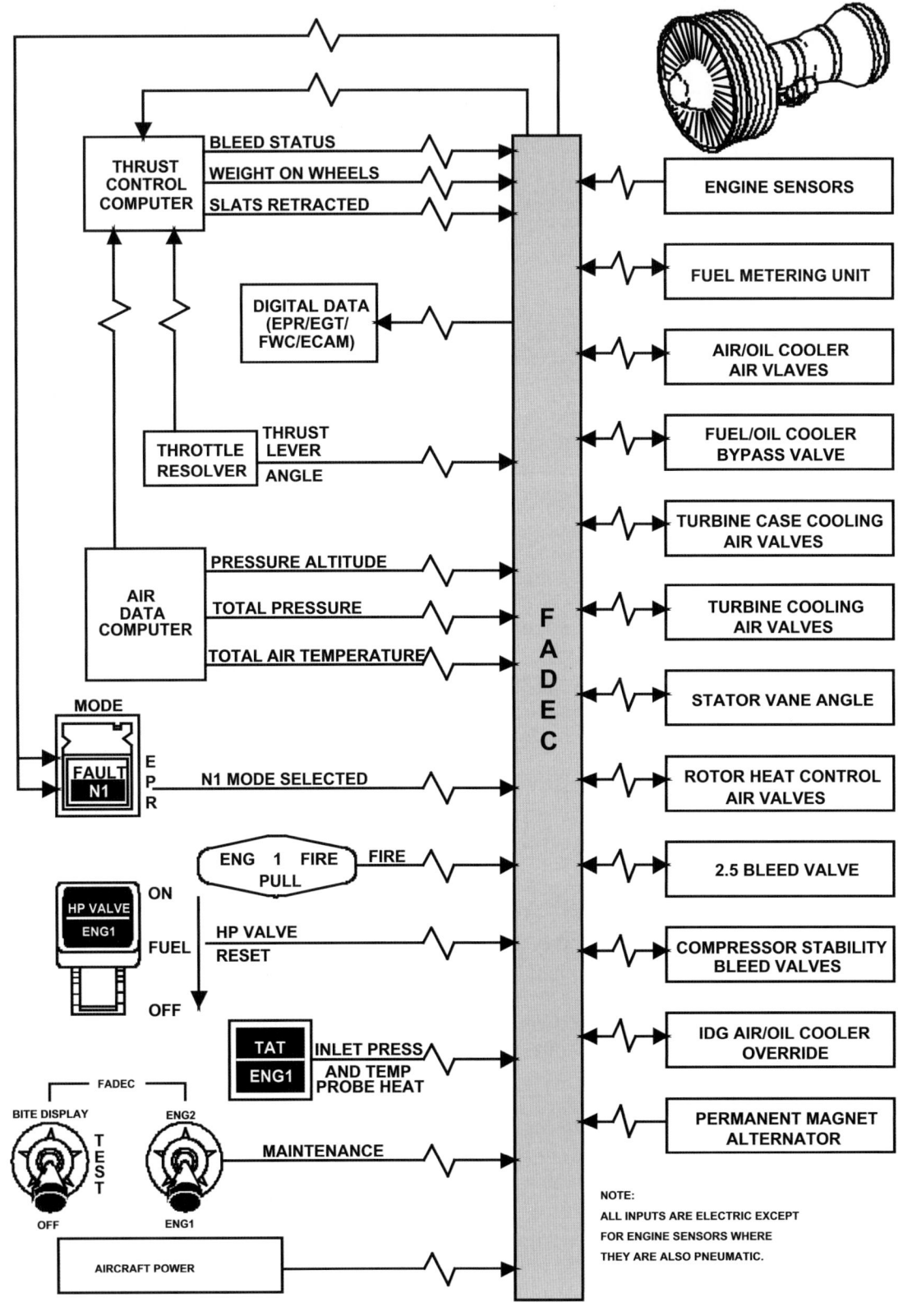

Fig. 23.1

Figure 23.1 indicates the signals that are transmitted between engine-mounted components and describes the engine/aircraft interface. The control has dual electronic channels, each with its own processor, power supply, programme memory, selected input sensors, and output actuators.

The FADEC has many advantages over the mechanical system.

> ➢ The control requires no engine adjustment, therefore no engine running, which saves fuel.
> ➢ The control reduces fuel consumption through improved engine bleed air control.
> ➢ The control fully modulates the active clearance control systems, producing a substantial benefit in performance by reducing engine blade tip clearances.
> ➢ The idle speed remains constant regardless of changes in ambient conditions and bleed requirements. In mechanical systems, the engine speed changes with ambient conditions.
> ➢ The higher precision of the digital computer ensures more repeatable engine transients (i.e. acceleration — deceleration) than those possible with hydro-mechanical systems. The latter is subject to manufacturing tolerances, deterioration, and wear that affect its ability to consistently provide the same acceleration and deceleration times.
> ➢ The control ensures improved engine starts by means of digital schedules and logic that adjusts for measured conditions.
> ➢ The control provides engine limit protection by automatic limiting of critical engine pressures and speeds. Direct control of the rating parameter also prevents inadvertent overboost of the selected rating during power setting.

The FADEC mounts on the engine compressor casing on anti-vibration mounts and is air-cooled.

A dedicated engine gearbox-driven alternator provides power to each electronic control channel. If computational capability is lost in the primary channel, the FADEC switches to the secondary channel. If a sensor is lost in the primary channel, crosstalk with the secondary channel supplies the information.

In the unlikely event of the loss of both channels of the electronic control, the torque motors are spring loaded to the following failsafe positions:

> ➢ Fuel flow goes to minimum flow
> ➢ Stator vanes are set fully open (to protect take-off)
> ➢ The air/oil cooler goes to wide-open
> ➢ The active clearance control is shut off.

ENGINE CONTROL LIMITERS (AMPLIFIERS)

The engine control amplifier receives signals of exhaust gas temperature and engine speed (N_1 and N_2). The amplifier compares these parameters with pre-set datums. If either of these parameters exceeds their datums, a command signal goes to the HP pump to decrease output (this is achieved by opening the spill valve on the gear-type pumps, or reducing the swash plate angle on multi-plunger type pumps). This overrides the fuel control until the input condition has altered. The system is designed to protect against parameters exceeding their design values under normal operation and basic fuel system failure.

Chapter 24
Gas Turbine Engine Performance

STATIC THRUST

Static thrust is the product of mass airflow through the engine and rate of acceleration of the mass of air with the aeroplane stationary. The following formula applies:

$$T = M (V_2 - V_1)$$

T = thrust in pounds or newtons
M = mass of airflow in lb/sec or kg/sec
V_1 = initial velocity of a mass of air in ft/sec or m/sec
V_2 = final velocity of a mass of air in ft/sec or m/sec

ENGINE THRUST IN FLIGHT

Calculate the thrust generated in flight as follows:

$$T = MV_j$$

T = thrust in pounds or newtons
M = mass of air passing through engine in lb/sec or kg/sec
V_j = jet velocity at propelling nozzle in ft/sec or m/sec

Note: The mass of air (M) can also be written as the weight of air (W) divided by the gravitational constant (g) as 32.2 ft/sec^2 or 9.81 m/sec^2.

When the air passes into the engine it creates drag called **momentum drag**.

$$Momentum\ Drag = MV$$

M = mass of air passing through engine in lb/sec or kg/sec
V = aircraft speed in ft/sec or m/sec (TAS)

Momentum drag must be deducted from the thrust to find the actual forward force, which is **net thrust**. To calculate the net thrust combine the above formulas:

$$Net\ Thrust = M (V_j - V)$$

For a turbofan engine, the two flows of the bypass and the core must be considered. The resulting formula is:

$$\text{Thrust} = \text{Bypass M} (V_j - V) + \text{Core M} (V_j - V)$$

The idle values of rpm and thrust are approximately 25% N_1 and 5% of take-off thrust.

THRUST AND SHAFT HORSEPOWER

The performance of a turbojet engine is measured in thrust produced at the propelling nozzle or nozzles. Performance of the turbo-propeller engine is measured in **shaft horsepower (SHP)** produced at the propeller shaft. However, both types are mainly assessed on the amount of thrust or SHP they develop for a given weight, fuel consumption, and frontal area.

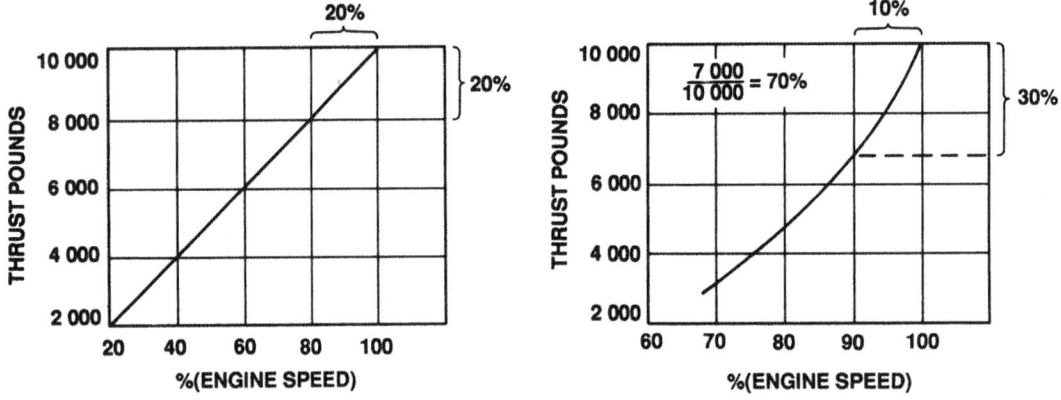

Fig. 24.1

The thrust or SHP developed depends on the mass of air entering the engine and the acceleration given to it during the engine cycle. For example:

$$\text{Thrust} = \text{Mass [of air]} \times \text{Acceleration [of air]}$$

It is obvious that such variables as the aeroplane forward speed, altitude, and climatic conditions influence the value of this thrust or SHP.

The available thrust is, however, limited by the turbine inlet temperature, which must not be exceeded because of the materials used and turbine assembly design. Improved materials and more efficient turbine cooling have been developed over the years. These have led to an increase in the turbine operating temperature.

VARIATIONS OF THRUST WITH SPEED, TEMPERATURE, AND ALTITUDE

There are a number of conditions that affect the performance of gas turbine engines. In general, if a specific amount of fuel is supplied to the engine, the thrust of the engine varies depending on the temperature and pressure of the air that enters the air inlet.

SPEED

As the forward speed increases, the thrust will reduce due to a combination of:

> ➢ Inlet momentum drag
> ➢ Decreased acceleration of the airflow (i.e. the jet velocity remaining relatively constant with increasing inlet velocity)

Due to the **ram effect** obtained from increasing forward speed, additional air is forced into the engine, increasing the mass airflow and air velocity. The effect of these increases tends to offset the increased inlet momentum drag that occurs with increased forward speed. The resultant decrease in net thrust partially recovers as aeroplane speed increases.

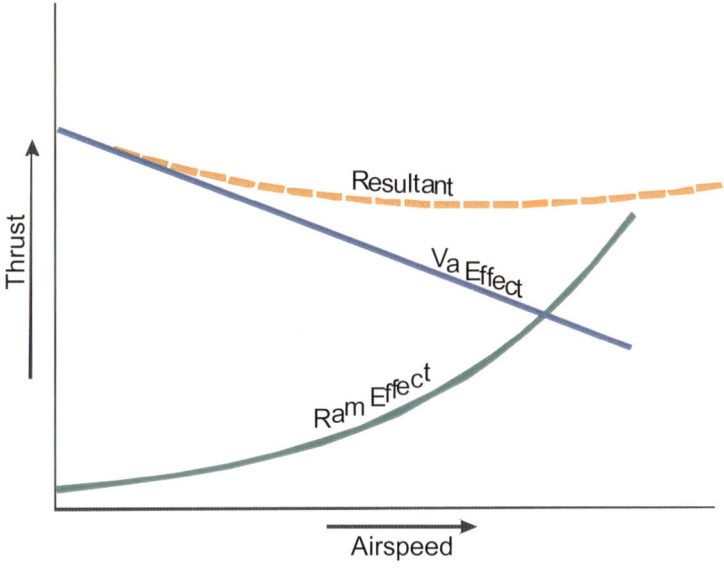

Fig. 24.2

Thus, **ram effect** is of great importance to gas turbine engine performance, especially at high speed.

TEMPERATURE

Cold air increases the density of the air, resulting in an increased mass of air entering the compressor for a specific engine speed, therefore increasing the thrust or SHP. However, the increased density of the air requires more power to drive the compressor. To maintain the same rpm, the fuel flow must increase; otherwise a fall in rpm occurs. Therefore, with a reduction in air inlet temperature, the engine either:

> ➢ Runs at reduced rpm but maintains the thrust
> ➢ If rpm is maintained constant, there exists an increase in thrust

Alternatively, hot air decreases the density of the air, which results in a reduction of the air entering the compressor, and the reverse occurs. When encountering high temperatures of typically 45°C, up to a 20% thrust loss can occur. In this situation, it may be necessary to employ some form of thrust augmentation (e.g. water or water/methanol injection). This is described later.

ALTITUDE

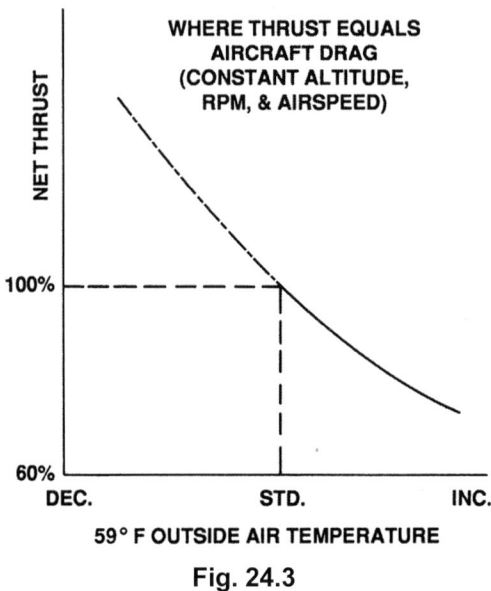

Fig. 24.3

As altitude increases, the ambient air pressure and temperature decrease. For a given engine speed this has the following effects:

>   The decreased atmospheric pressure reduces the density and hence the mass airflow into the engine, causing the thrust or shaft horsepower to fall.
>   The reduction in ambient temperature at altitude results in an increase in air density. This partially offsets the reduction in thrust caused by the fall in atmospheric pressure.

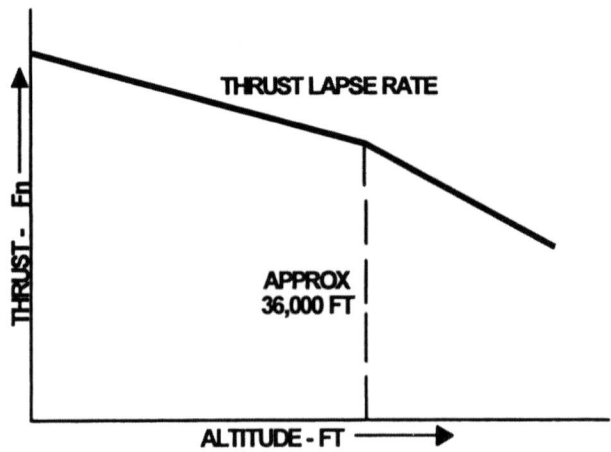

Fig. 24.4

At an altitude of 36 089 ft the temperature of the atmosphere is -56.5°C. Above this altitude, the temperature of the air remains constant until reaching altitudes above 65 617 ft. When an altitude of 36 089 ft is reached, the rate of decrease in thrust is greater. This is because the counteracting effect of the temperature decrease no longer balances the effects of decreasing pressure.

ENGINE PRESSURE RATIO (EPR)

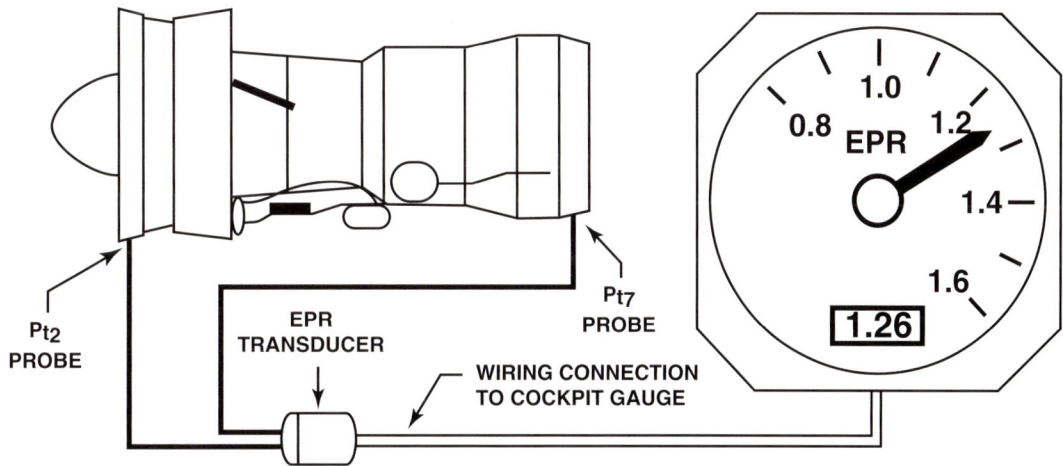

Fig. 24.5

Engine pressure ratio (EPR) is the ratio between the exhaust pressure and the compressor inlet pressure. It can be measured in a number of ways; turbine discharge pressure to compressor inlet pressure, on a fan engine it may be fan outlet pressure to compressor inlet pressure or integrated fan outlet/turbine discharge pressures to compressor inlet pressure.

ENGINE THRUST RATING

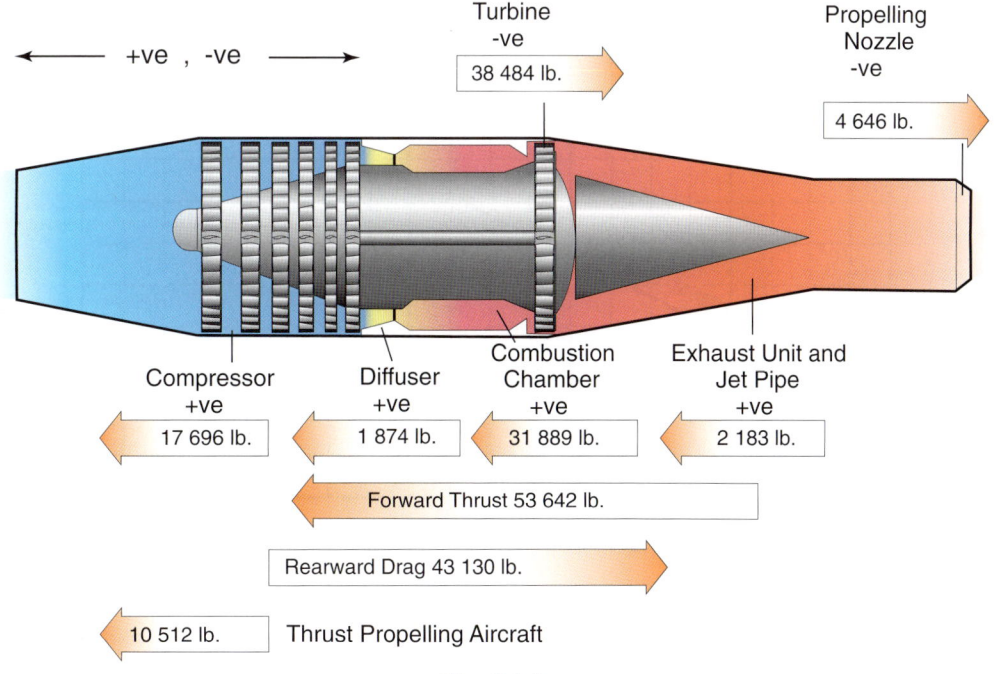

Fig. 24.6

An engine is rated by the amount of thrust in pounds or newtons it can develop.

FLAT RATED POWER

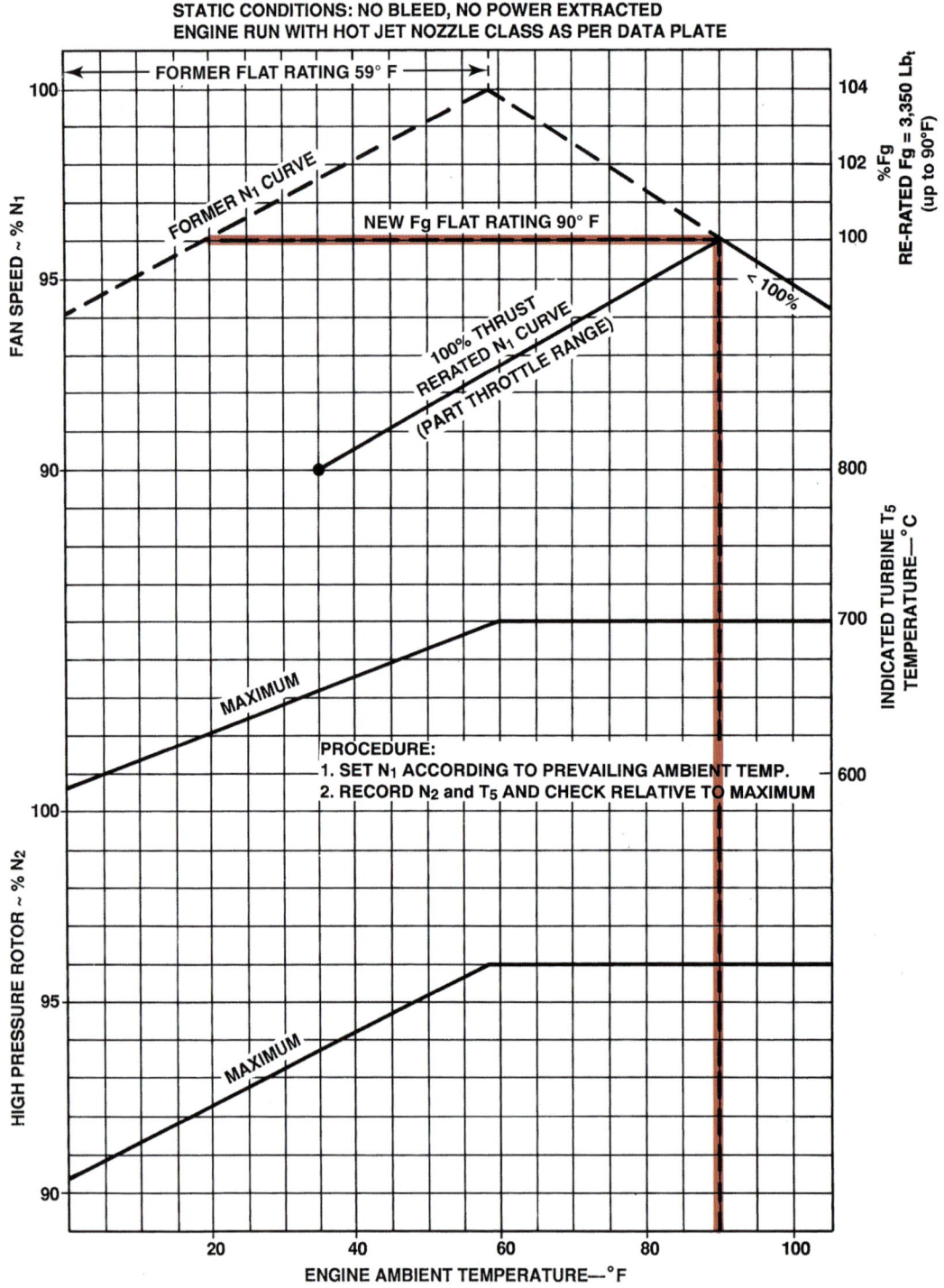

Fig. 24.7

Flat rated power means that the power output is restricted in cold ambient conditions, and is therefore able to give constant predictable power up to a specific limit (e.g. 29.9°C). The fuel control system prevents the limitations of shaft speeds, internal pressures, and turbine temperatures being exceeded should the specified limiting temperature be exceeded, by reducing fuel flow and power.

BLEED AIR

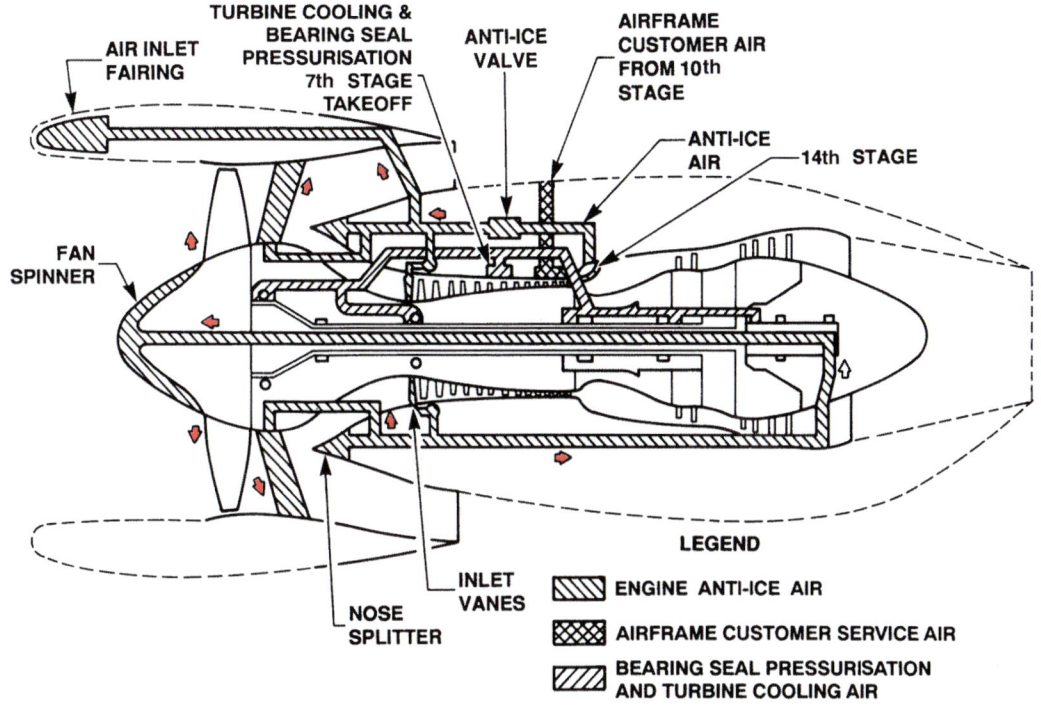

Fig. 24.8

Bleed air comes from the engine compressor to supply both internal and external requirements.

INTERNAL SUPPLIES

The heat, transferred from the main gas stream to the nozzle guide vanes, turbine blades, turbine discs, the bearings of the rotating assemblies, and the engine main casings, is absorbed and dispersed by directing a flow of comparatively cool air over these components. Due to the high temperature of the gas stream at the turbine inlet, it is necessary to provide internal air cooling of the nozzle guide vanes and the turbine blades. Air from the compressors also seals the bearing housings, thus preventing oil leakage into the main engine casings or into the compressor inlet. Compressor bleed air can be used as engine anti-icing of the nose cowl leading edge, inlet struts, nose cone, and inlet guide vanes. It may be necessary to provide accessory cooling during ground operations on items cooled in flight by ram air such as pumps, motors, and generators.

EXTERNAL SUPPLIES

There are aeroplane systems that require engine bleed air to operate (e.g. hydraulic pumps, generators, and motors). Such systems may involve the supply of an emergency system or the operation of certain high-lift devices. As gas turbine engines usually fly at much higher altitudes where pressurisation and air conditioning are necessary, they require another form of engine take off. This involves tapping high-pressure air from various stages of the compressor. This air is not only for air conditioning and pressurisation but can serve for anti-icing systems, hydraulic header tanks, and fuel tank pressurisation.

Under certain circumstances, cabin bleed air must be closed, (e.g. to prevent smoke and fumes entering the cabin) especially on some smaller engines during phases of flight where the removal of compressor air is critical to thrust. In some aeroplanes, compressor air serves for self-contained engine starting systems and for boundary layer control for take-off and landing.

EFFECTS OF BLEED AIR EXTRACTION

The bleeding of air from the compressor reduces the amount of mass airflow, therefore decreasing thrust/rpm, and increasing exhaust gas temperature and specific fuel consumption.

THRUST AUGMENTATION

In certain instances, it may be necessary to recover or improve thrust being developed by an engine. A **thrust augmentation system** achieves this. The principal methods are:

> ➢ Afterburning or reheat
> ➢ Water or water/methanol injection

AFTERBURNING

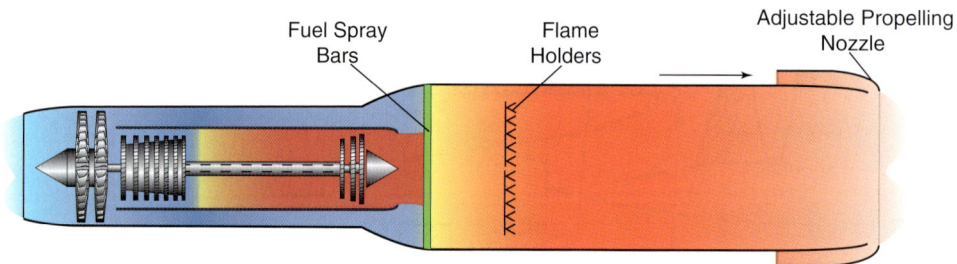

Fig. 24.9

Afterburning is a method of augmenting the basic thrust of an engine to improve take-off, climb, and acceleration of the aeroplane. This increase in thrust is obtainable using a larger engine but is wasteful in terms of weight, drag, and SFC. Afterburning is a useful method of thrust augmentation for short periods.

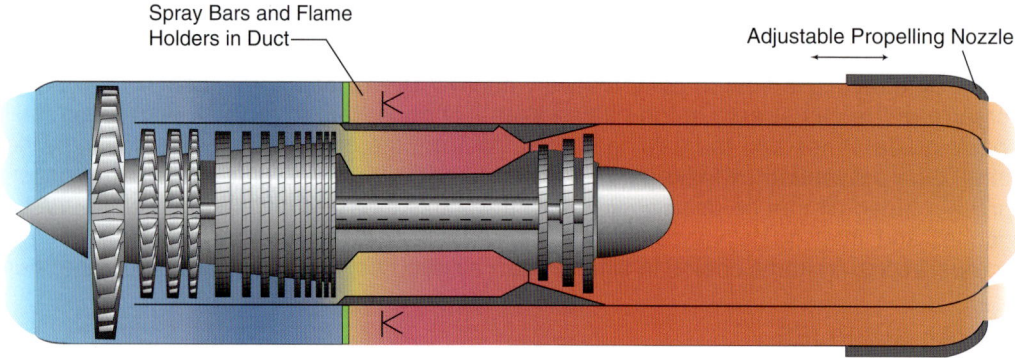

Fig. 24.10

The principle of afterburning is to introduce fuel between the turbine and propelling nozzle, utilising the un-burnt oxygen in the exhaust stream to support combustion. The resultant temperature increase provides an increased velocity to the jet efflux leaving the propelling nozzle, thus increasing thrust. The thrust increase can range from approximately 30% for a turbojet up to 70% for a low by-pass engine.

AFTERBURNING SYSTEM

Afterburning temperatures are very high (1700°C). As a result, fuel is introduced via a special manifold so that the normal exhaust gas can insulate the jet pipe walls. The propelling nozzle area is variable, either two-position or infinitely variable, to prevent back pressure affecting the normal engine operation and making afterburning over a wide speed range possible.

Ignition is not spontaneous and is usually initiated by catalytic igniter, igniter plug, or **hot shot** which is a flame streak from the combustion chamber.

To co-ordinate fuel flow/nozzle area so that the pilot can select varying degrees of reheat and still maintain the correct relationship between these two requires a control system. Control is achieved via a pressure ratio sensing device (compressor outlet to jet pipe pressure). This ensures that the engine pressure ratio remains unaffected by afterburner selection.

Afterburning was, for a long time, confined to military aeroplanes, but with the advent of the Concorde and the Tu144, with their need for rapid transonic acceleration, reheat became available for civil transport. Although SFC increases very much during use, the improved climb rate, rapid acceleration, and power reserve at take-off more than compensates.

However, the cost of aviation fuel has risen since the conception of supersonic passenger travel, making it unlikely in the foreseeable future for supersonic mass passenger travel to be reinstated. Conversely, with the advent of the more fuel-efficient turbofan engines, aircraft have actually slowed down but carry larger numbers.

WATER INJECTION

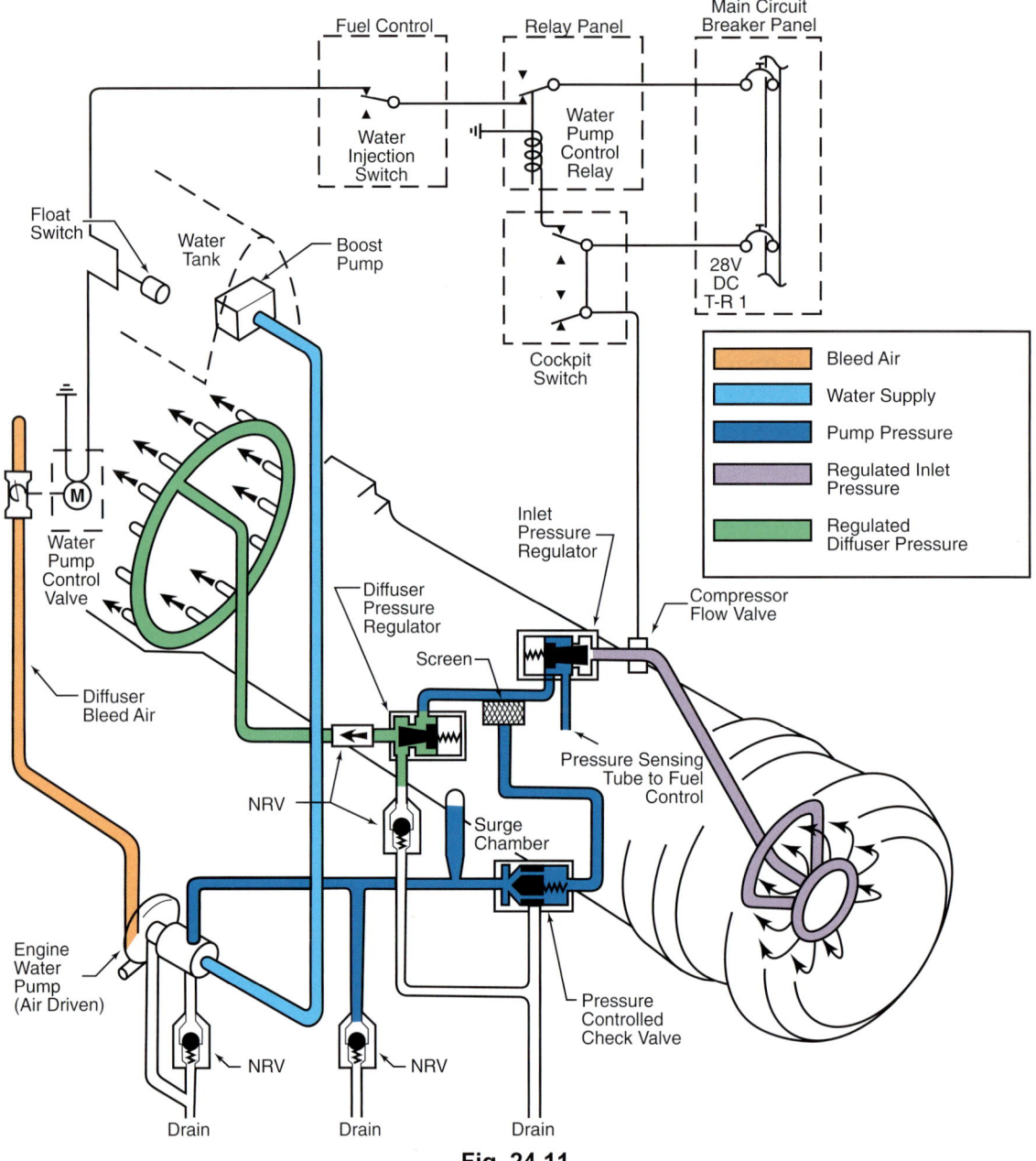

Fig. 24.11

The density of the airflow passing through the mechanism affects the power output of a gas turbine. As a result, a reduction in thrust or shaft horsepower results when density decreases due to an increase in ambient air temperature or altitude. Under these circumstances, the power output can be restored, or can be boosted to a value over 100% max power, by the injection of a water-methanol mixture at the compressor inlet, or water at the combustion chamber inlet.

When sprayed directly into the compressor inlet, air temperature is reduced, resulting in an increase in air density and thrust. Injecting water only would reduce the turbine inlet temperature. However, since methanol is added, burning it in the combustion chamber restores the turbine inlet temperature, resulting in power restoration without having to adjust the fuel flow.

Spraying coolant into the combustion chamber inlet results in an increase in the mass flow through the turbine relative to the mass flow through the compressor. As a result, the pressure and temperature reduction across the turbine decreases, resulting in an increased jet pipe pressure that produces additional thrust. As a result of water injection, a reduction in turbine inlet temperature occurs and the fuel system schedules an increase of fuel flow to increase the maximum rpm, which results in extra thrust. Again, if methanol is used, burning the methanol in the combustion chamber partially or fully restores the turbine inlet temperature.

SYSTEM OPERATION

Usually it is a requirement to switch on the system and coolant delivery from a tank via a pump controlled by the following:

- ➢ Throttle advancement to the take-off position
- ➢ Engine parameters
- ➢ Atmospheric conditions

The coolant is stored in an aircraft tank and pumped to a control unit that meters the flow of the coolant. System operation is usually indicated on the flight deck.

Chapter 25
Powerplant Operation and Monitoring

INTRODUCTION

Engines are rated to cover all aspects of operation, including:

Maximum take-off thrust the maximum thrust certified for take-off, normally limited to five minutes.

Maximum go-around thrust the maximum permissible thrust during go-around.

Maximum continuous thrust the maximum thrust certified for continuous use.

Maximum climb thrust the maximum thrust approved for normal climb operation.

Maximum cruise thrust the maximum thrust approved for normal cruise operation.

TAKE-OFF

When cleared for take-off, advance the throttle lever in a smooth and unhesitating way to the required take-off thrust position. As the throttle advances, monitor the instruments to ensure that the engine is functioning properly. Obtain high thrust whilst the aircraft is stationary or soon after the aircraft starts to roll. In this way, the thrust stabilises well before the aircraft takes off. Once the throttle is set to take-off thrust, no further adjustments are necessary until airborne. Throughout the take-off, the engine instruments must be closely monitored and action taken to ensure that engine limitations are not exceeded. At take-off thrust, an engine is operating closer to its all-out physical and structural capabilities than during any other phase of its operation. Internal operating temperatures, more than anything else, affect the service life of turbojet engines. Abnormally high temperatures shorten the life of turbine nozzles, discs, and blades. Therefore, permissible take-off thrust is limited to no more than five minutes. When long runways are available, make reduced thrust take-offs by reducing the amount of thrust by a few percent. This extends engine life.

CLIMB

When climbing at a fixed throttle setting, the temperature of the outside air decreases and the fan speed tends to increase. Normally, only one or two throttle adjustments are necessary throughout the climb, depending on whether performing a high-speed or long-range climb. Constantly monitor exhaust gas temperatures (EGT) to stay within engine operating limits. Also monitor fuel flow, since fuel flow provides a good check on proper engine operation. One of the first signs of engine malfunctions or fuel control problems is abnormal or erratic fuel flow.

CRUISE

Once thrust is set to obtain the desired cruising speed, the throttle may remain fixed throughout the cruise. The amount of fuel consumed during a short flight represents only a slight decrease in aircraft weight. On longer flights where the burning of fuel results in substantial decrease in aircraft weight compared with take-off weight, the speed of the aircraft tends to increase. However, maintaining constant cruise speed attains optimal economy and operational efficiency. Therefore, periodically reduce the thrust of the engine, which results in lower fuel costs and increased engine life.

DESCENT

Standard descent procedures for aircraft powered with turbojet engines require relatively high speeds and rates of descent. This reduces stress on the engine and allows for quick action during an emergency. During the initial part of the descent, the throttle usually remains in the cruise position. When at lower altitudes, retard the throttle smoothly and slowly (if conditions permit). Slow throttle movements reduce rapid temperature changes in the engine and allow regulating systems in the engine to respond fully.

APPROACH AND LANDING

During approach and landing, rapid engine response to throttle movements might be required. The use of variable stator vanes allows this rapid response and ensures stall-free operation. It is good operating technique to keep engine speeds as high as is practical throughout the approach to reduce engine response time when quick thrust changes are needed. In the event that thrust must be applied for a go around manoeuvre, advance the throttle part way and hold momentarily in an intermediate position to ensure proper engine response. Then move the throttle rapidly to the go-around thrust position.

ENGINE IDLE RPM

There are two engine idle values, **ground idle** and **flight Idle**. In some installations, the terms **minimum idle** and **approach idle** are used. Ground idle is the engine speed, typically for use in ground operation of a gas turbine engine to produce the minimum amount of thrust. However, on some installations it can be used for all phases of flight except when anti-icing is ON or during approach and landing. Flight idle is the lowest recommended operating speed in flight and is a higher value than ground idle to enable the engine to produce a short acceleration time, which may be needed in the case of a go around, and also to compensate for a reduction of airflow through the engine due to engine and aeroplane bleed air requirements.

CONTROL OF THRUST/POWER

The control of a gas turbine engine is usually by one control lever. The control lever might be called a throttle, power, or thrust lever. Operation of the control lever selects the required thrust level and the fuel control system automatically maintains this level. In some autothrottle systems, with changes in thrust a resulting movement of the thrust lever can be observed. On modern aeroplanes, the thrust lever is integrated with the FADEC, autothrottle/thrust, and the thrust management computer.

On a FADEC equipped aeroplane, set the thrust by positioning the thrust lever angle to align the control EPR command with the thrust management computer reference. Once positioned, the throttle does not move, as the control maintains the preset thrust value. Automatic acceleration and deceleration to maintain EPR may result from changes in flight or environmental conditions. EPR is maintained until the thrust lever moves to a new setting. Thrust lever movement on a FADEC is manual in operation when auto-thrust is engaged; any thrust changes required are made without the thrust lever moving.

Turbo-propeller engines can have an interconnection of the throttle with a propeller control unit, therefore ensuring single-lever engine operation that controls both fuel flow and rpm, with the overall controlling factor being the fuel flow to prevent exceeding engine operating limitations. Some turbo-propeller engines are constant speed and incorporate a condition lever that selects an rpm for ground (feather/fuel off and low for taxi) and flight conditions (high for take-off and cruise) and once set it only requires resetting when flight conditions change. There are two operating mode ranges for a turbo prop, these are:

> ➢ **Alpha Range** — the flight operational mode of a turboprop engine, including all operations in the flight range from take-off to landing (i.e. flight fine pitch to feather).

> ➢ **Beta Range** — the engine ground operational mode when the flight deck control lever between ground fine pitch and reverse thrust hydro-mechanically controls the propeller pitch.

ENGINE MONITORING

When installing a turbojet, turbofan, or turboprop engine in an aeroplane, various instruments, controls, and warning devices are necessary for normal control and operation of the engine and display either conventionally or electronically. Most aeroplanes have the following instrumentation.

ENGINE SPEED (RPM)

A speed indicator is an instrument that indicates the speed of an engine and is calibrated in percent of a designated maximum number of revolutions per minute (rpm) rather than actual rpm. Normally the rpm of all compressors are indicated in terms of N, as previously described in the Compressor chapter.

Fig. 25.1

On multi-spool engines, the HP compressor speed indicator is primarily referred to during engine start, and on a high-bypass turbofan engine, the fan speed provides an accurate indication of engine thrust.

Speed indication can either be by an engine driven tacho-generator that transmits electrical signals to the indicator, or a variable-reluctance probe in conjunction with a phonic wheel or fan blades that induce an amplified electrical current that is transmitted to the indicator. The latter system dispenses with the need of a tacho-generator, thereby reducing the amount of moving parts and weight, as illustrated below.

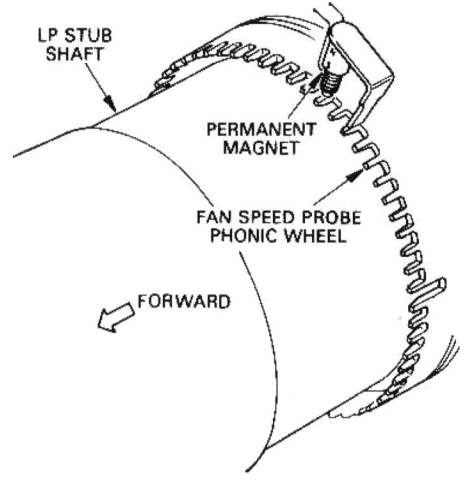

Fig. 25.2

ENGINE PRESSURE RATIO INDICATOR

An engine pressure ratio (EPR) indicating system, used in conjunction with the speed indication systems, allows monitoring of the engine performance to obtain the required amount of thrust. This system is only used on some engine types.

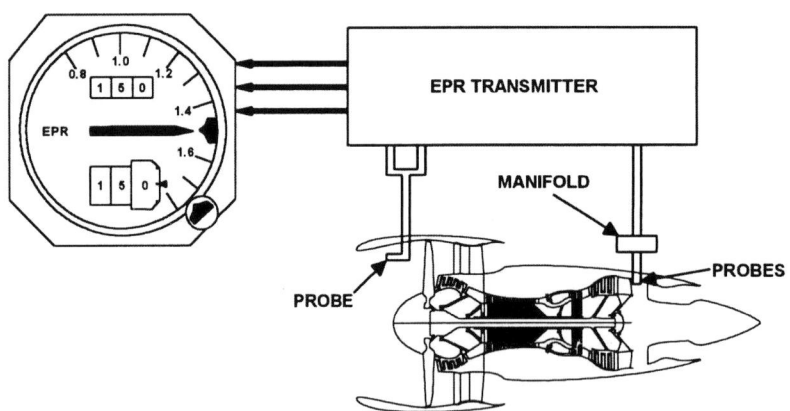

Fig. 25.3

TURBINE GAS TEMPERATURE

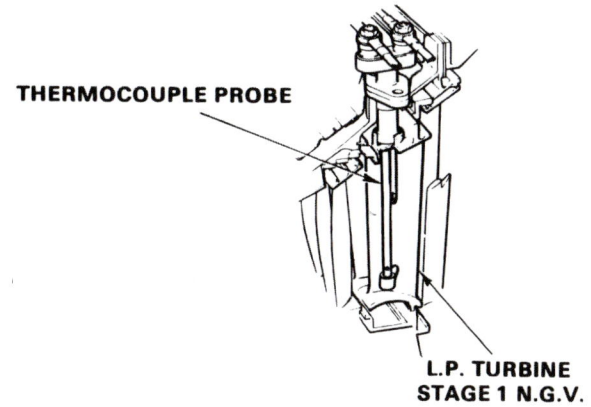

Fig. 25.4

The efficiency and durability of a gas turbine engine directly relates to the temperatures to which the high-pressure turbine is subjected. The exhaust gas temperature indicator displays the exhaust gas temperature in degrees Celsius and is the average of the temperatures measured by several thermocouple probes located either at the exhaust unit (EGT), jet pipe (JPT), or within the turbine at one of the stator positions (turbine gas temperature, turbine entry temperature, turbine inlet temperature).

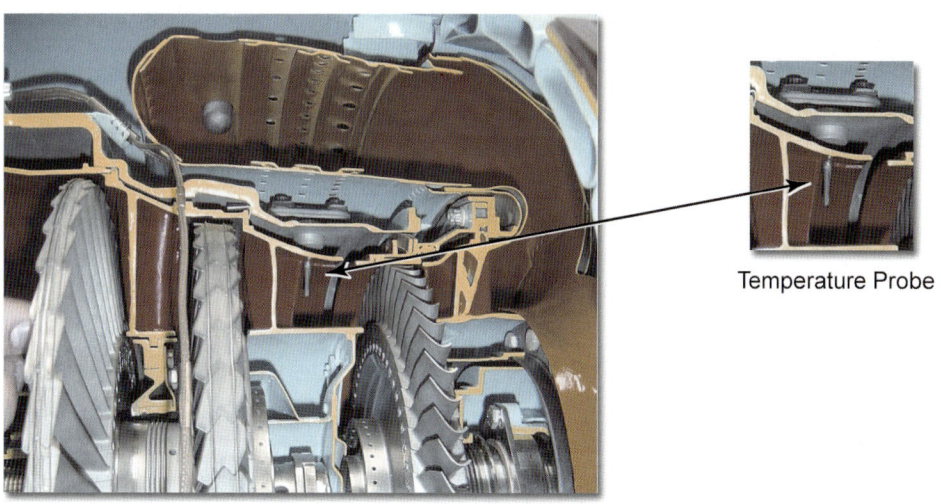

Fig. 25.5

The probe operates on the thermocouple principle, the probe being the hot junction and the instrument being the cold junction. Heating results in a current flow, therefore these probes operate independently of the aircraft's electrical system. The probes can be single, double, or triple element to give a more accurate indication, connect in parallel to give an average reading, and are not affected should a probe or probes fail.

OIL TEMPERATURE AND PRESSURE

Fig. 25.6

Oil temperature is measured by a sensitive element in the oil system, and indicates in degrees Celsius. Variations within the engine are quickly noted and engine performance can be inferred.

Oil pressure is measured at the outlet of the pressure pump, and unexpected pressure variations during engine operation may indicate lubrication malfunctions. Additionally, a low-pressure switch in the lubrication system illuminates a low pressure warning light on the flight deck.

FUEL TEMPERATURE AND PRESSURE

The temperature and pressure of the fuel supply are electrically transmitted to indicators on the flight deck and are similar in operation to the oil system indicators. A fuel differential pressure switch is fitted on some engines to the LP fuel filter that senses pressure differential across the filter element and is connected to a warning lamp providing indication of impending filter blockage and possible fuel starvation.

VIBRATION

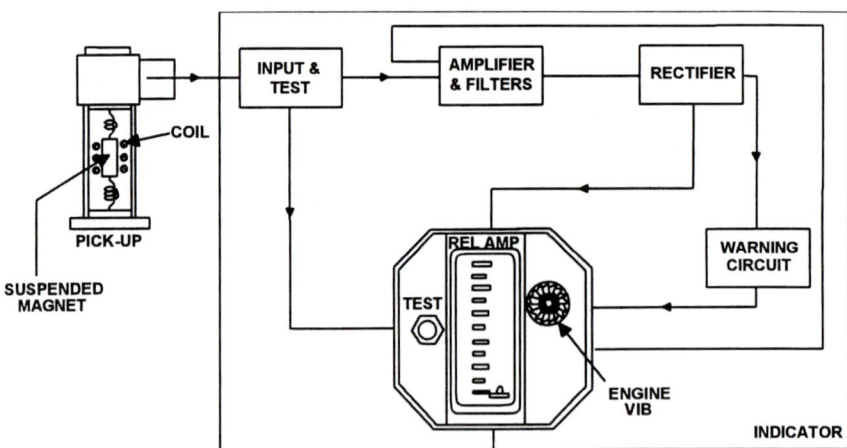

Fig. 25.7

A vibration indicator indicates the amount of engine vibration, providing information about the overall mechanical performance of the engine.

Relative amplitude indicates vibration and if detecting an unacceptable level of vibration, a warning light illuminates on the flight deck. There is also a red line warning on the indicator.

Engine-mounted transducers monitor vibration. These can be either electro-magnetic or piezoelectric design. They convert vibration rates into electrical signals that result in the pointer of the indicator moving in proportion to the level of vibration, which is proportional to the amount of rotor imbalance. The signals are amplified, electronically filtered, and sometimes selectable between frequency ranges.

ENGINE TORQUE

Fig. 25.8

On turboprop engines, a torquemeter is used in indicating power where the torque produced at the propeller shaft is usually measured, since jet thrust is only a small proportion of the engine power. One type of torquemeter uses helical gear teeth in the reduction gear and consequently an axial thrust develops by layshafts, which is proportional to the power transmitted through the reduction gear. An opposing oil pressure, proportional to engine power, balances this axial thrust and is called torquemeter pressure, indicated as pounds per square inch (psi) via a transmitter on a flight deck gauge. On some installations, the torquemeter pressure may be applied directly to the water/methanol and automatic feathering systems.

ELECTRONIC INDICATING SYSTEMS

Modern day aeroplanes use the glass cockpit in the form of electronic displays to replace the conventional instruments. There are two systems currently in use:

> ➢ Engine Indicating and Crew Alerting System (EICAS)
> ➢ Electronic Centralised Aircraft Monitor (ECAM)

EICAS

This system consists of two display units, one control panel, and two computers, supplied with analogue and digital signals. Only one computer controls, whilst the other is on standby. Should a failure occur the standby computer can turn on either automatically or manually. The displays are cathode ray tubes (CRT) or LCDs and are mounted one above the other.

The upper display is the primary display and displays primary engine parameters (e.g. N_1, EGT, and in some installations EPR). It also displays warning and caution messages. The primary engine parameters are permanently displayed in flight.

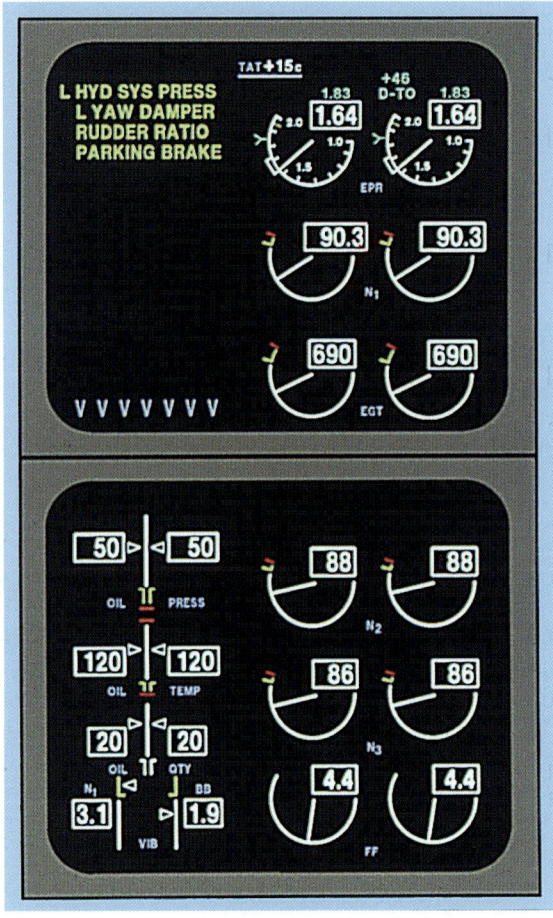

Fig. 25.9

The lower display is the secondary display and displays secondary engine parameters such as N_2, N_3 (which is applicable to some Rolls Royce engines), fuel flow, oil quantity, oil pressure, oil temperature, and engine vibration, plus non-engine systems status (e.g. hydraulic system, electrical system, etc.). The secondary display is normally blank in flight, but is selected to indicate secondary engine parameters during start.

Should a display fail, the information automatically transfers to the other screen in a format called compact. If the total EICAS display is lost, a standby LCD engine indicator provides primary engine information.

ECAM

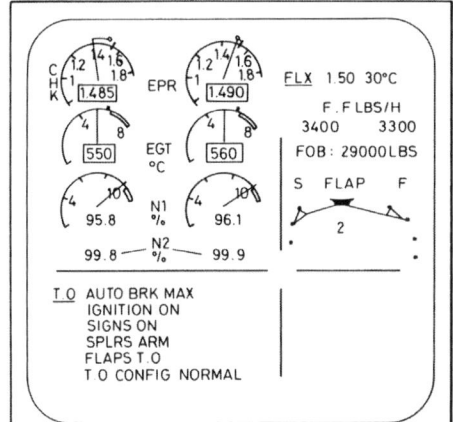 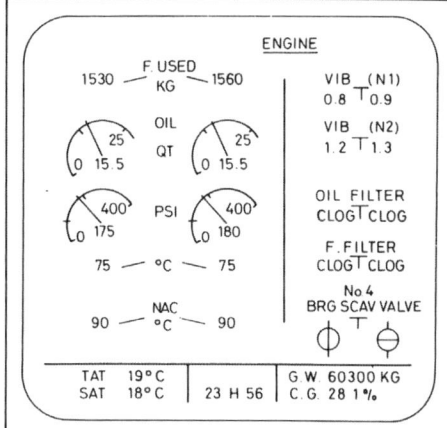

Fig. 25.10

This system was originally developed for the Airbus, and has the same basic components as the EICAS system. The processing and display of information differs quite significantly in that it displays in a checklist and a pictorial or synoptic format. Depending on the aeroplane, the displays can either be mounted one above the other or side-by-side.

The upper or left display is the engine and warning display and displays engine parameters, status of systems, warnings, and corrective action in a sequenced checklist format.

The lower or right display is the Systems Status Display and displays associated information in a pictorial or synoptic format.

WARNING SYSTEMS

These are provided to give an indication of a possible failure or of a dangerous condition that exists, so the crew can take action to ensure the integrity of the engine or aeroplane. In the case of electronic indicating systems, the severity of the warning dictates the colour displayed.

Chapter 26
Auxiliary Power Unit (APU) and Ram Air Turbine (RAT)

AUXILIARY POWER UNIT (APU)

Fig. 26.1

These units are fitted to provide a source of electrical power, pressurised air (air conditioning and main engine starting), and in some cases hydraulic power (via an integral pump) on the ground when the main engines are shut down. This makes the aeroplane less dependent on ground support equipment. In some instances, the APU is used in flight to provide emergency power, especially for ETOPS operations.

When used in flight, the maximum operating and maximum starting altitude parameters published in the flight manual must be adhered to. The maximum starting and operational heights vary from type to type but for modern aeroplanes can be as much as 43 000 ft and 45 000 ft, respectively. The minimum to maximum declared airspeed range of the declared relight envelope should cover at least 30 kt. Maintenance is similar to that used on the main aeroplane power units.

GENERAL DESCRIPTION

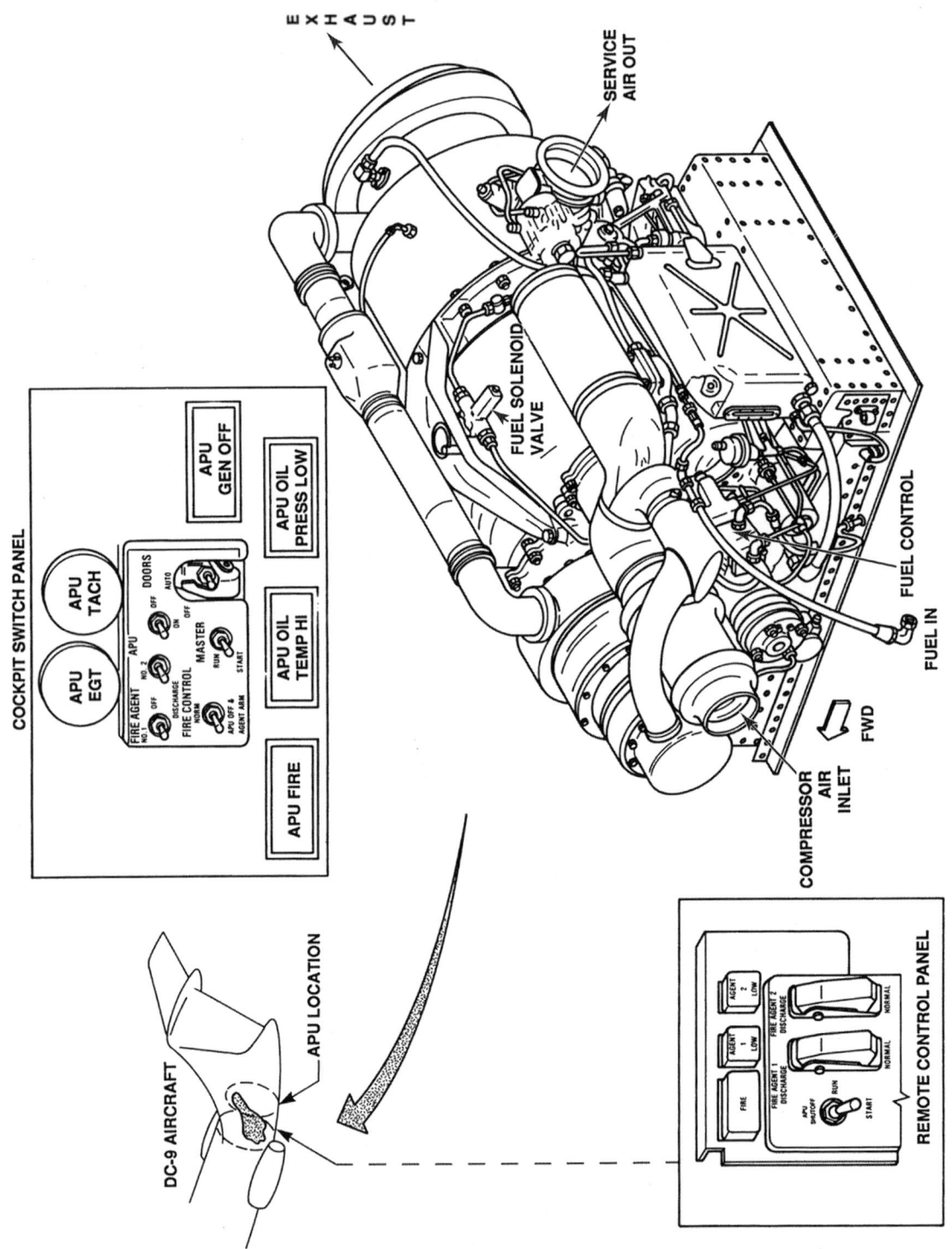

Fig. 26.2

The APU is a self-contained unit that normally consists of a small constant-speed gas turbine engine coupled to a gearbox. This gearbox drives a generator of a similar type and power rating to the main engine-driven generators. This gearbox also drives the APU accessories, such as fuel pump, oil pump, tachometer generator, and a centrifugal switch.

The purpose of the centrifugal switch is to control the starting and ignition circuits, the governed speed indication circuit, and the overspeed protection circuit of the APU.

LOCATION

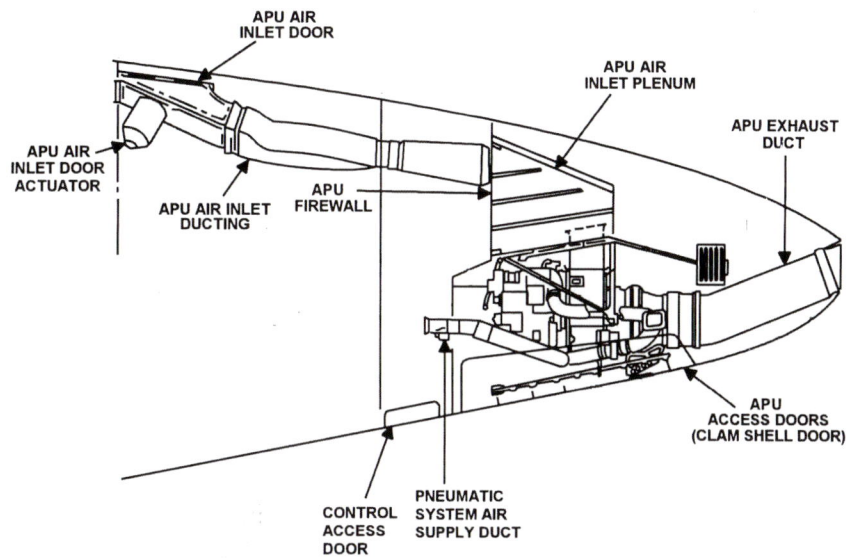

Fig. 26.3

The APU is normally located in an unpressurised compartment of the fuselage, usually in the tail section. This compartment is separated from the remainder of the fuselage by a firewall, and the unit is secured to the structure by rubber-bonded anti-vibration mountings. Access to the compartment is normally via hinged cowling panels.

Fig. 26.4

AIR SUPPLY

Air for the APU compressor is drawn in through either single or twin intakes, connected via ducting to the intake section. Doors provided in these intake sections usually open and close by electrical actuators. These actuator circuits, interconnected with the APU master control switch, ensure the correct operating sequence during starting and shutdown. Indicator lights, which, depending on the installation, connect to micro-switches or proximity switches at the door locations, indicate door positions.

The APU compressor discharges air not required for combustion into a plenum chamber, connected via ducting to the air conditioning system and the main engine air-starting system of the aeroplane. The air supply automatically regulates to provide the correct amount without overloading the APU.

FUEL SUPPLY

Fuel supply is from one of the tanks in the main fuel system via a solenoid-operated valve and is regulated by a fuel control unit that controls the acceleration of the APU and maintains the speed by proportioning fuel flow to load conditions.

LUBRICATION

A self-contained system, consisting of an oil tank, pump, filter, cooler, and oil jets, lubricates all the gears and bearings within the APU. Indicator lights monitor the system operation, as well as instruments associated with functions such as oil pressure, temperature, and quantity.

STARTING AND IGNITION

An electric starter motor connected to a drive shaft in the accessory gearbox rotates the engine for starting. Power for the starter motor is drawn from the aeroplane's batteries, its APU, or an external power source. The ignition system is of the high-energy type and is controlled from the master control switch.

Note: The APU can already be running, but not started, during refuelling operations.

COOLING

A fan driven by the APU accessory gearbox normally provides cooling and ventilation of the APU compartment. This air also ducts from the fan for cooling the AC generator and APU lubricating oil.

ANTI-ICING

In some APU installations, the air intake area is protected against ice formation by bleeding a supply of hot air from the compressor over the inlet surfaces.

FIRE DETECTION AND EXTINGUISHING

A continuous wire detection system and a single-shot fire extinguisher normally accomplish the detection and extinguishing of a fire in an APU compartment. The arrangement of detection circuits is so that, in addition to actuating warning systems, they automatically shut down the APU. The fire extinguisher bottles can be discharged manually or automatically.

CONTROLS AND INDICATORS

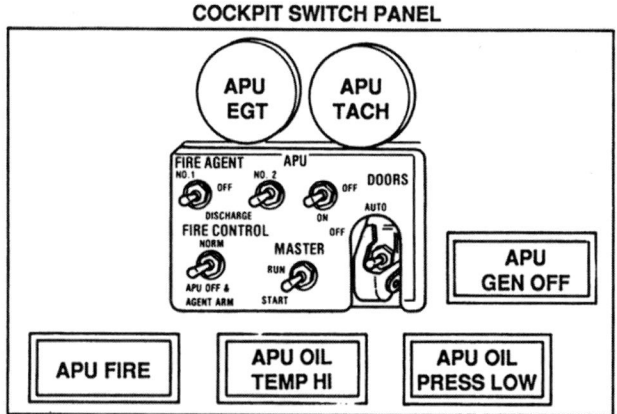

Fig. 26.5

All switches, warning lights, and indicating instruments necessary for the starting, stopping, and normal operation of the APU are located on the flight deck and in fuselage compartments accessible from outside the aeroplane. Normally, an APU can only be started from the flight deck, but can be shut down from either location.

Operation of the APU is monitored by an exhaust gas temperature indicating system and in most installations, a system to record the number of hours the APU has been in continuous operation. Depending on the installation, provisions for monitoring APU starting current, engine rpm, generator output voltage and frequency, generator bearing temperature, and connection of an APU test set may also be included.

APU SHUT DOWN

An APU is normally shut down by allowing it to operate at no-load governed speed for approximately 2 minutes, and then selecting the OFF or STOP position of the master control switch. Depending on the type of APU and its installation requirements, shut down of an APU can also take place automatically because of any one of the following conditions:

> ➤ High exhaust gas temperature
> ➤ Loss of exhaust gas temperature signal to the electronic control system
> ➤ Overspeed
> ➤ Low oil pressure
> ➤ High oil temperature
> ➤ Opening or closing of cooling air shut-off valve before 95% of governed speed has been attained
> ➤ Overheating of the APU bleed air delivery duct just forward of the APU compartment
> ➤ APU fire detection system operation
> ➤ When specified airspeed or altitude limitations are exceeded
> ➤ During take off operation of landing gear shock strut micro-switches

In some installations, the APU can also be shut down in an emergency by using a FIRE switch on the control panel, or by pulling a FIRE handle on the flight deck panel. Take care when using a FIRE switch, which arms the fire extinguisher discharge circuit, not to inadvertently discharge the extinguishant. If an automatic shut down has occurred, select the master switch to the OFF or STOP position.

RAM AIR TURBINE (RAT)

The RAT is used to supply the aeroplane with an emergency source of hydraulic power to the flight controls, etc, in the event that all systems fail, and normally stows in the underbelly fuselage. The RAT can be deployed in flight by manual selection at any time. However, if all the hydraulic systems' pressure drops, it deploys automatically. Ground sensors inhibit automatic deployment of the RAT on the ground.

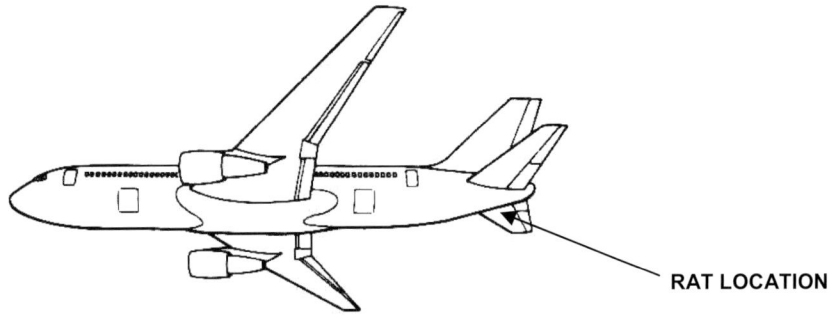

RAT LOCATION

Fig. 26.6

The RAT consists of a variable pitch propeller, driven by the airflow. Bob-weights and springs govern propeller speed, producing a constant speed. When initially deployed, the blades are in fine pitch allowing the propeller to spin up to the governed speed as quickly as possible. When at its governed speed of approximately 4000 rpm, the propeller blade pitch increases to prevent over-speeding.

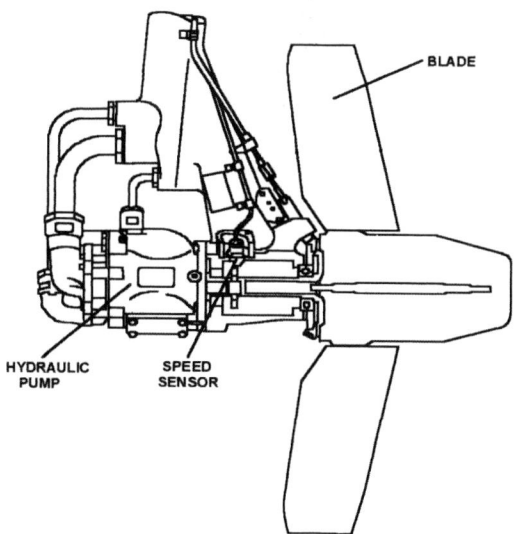

Fig. 26.7

A variable delivery hydraulic pump attaches directly to the output shaft of the propeller. After initial deployment, the pump is off-loaded by porting the pressure line back to the return line, allowing a pre-determined volume of fluid to refill the RATs cartridge. When the cartridge is full, the porting to the return line closes and all the fluid produced is directed to the aircraft's primary systems. The deployment and production of full system pressure is achieved in approximately 3 seconds. A RAT deployment light, which is normally amber or red, is near the RAT manual deployment switch. In addition, there is a green RAT pressure light to indicate that the system is up to pressure.